"Dr. Lombard's deep knowledge of the brain and its intricate mechanism helps us to unveil the mystery of the divine and the reality of our soul. This book is highly recommended to all who seek to know who they are, pursuing in an honest, simple, and open way, the modern and the ancient."

—Menas Kafatos, PhD, coauthor of the
New York Times bestseller *You Are the Universe*

"In the ever-raging battle between faith and science, the neurologist Jay Lombard is one of those rare emissaries who communicate in the language of both camps."

—*New York Times*

"Contextualizing scientific precepts with humanizing personal accounts of psychiatric patients—and family members—Lombard erects a much-needed bridge between science and faith."

—*Booklist,* starred review

"*The Mind of God* is an elegant and wise book that leads us to wonder about and fashion our own unique answers to some of the most baffling and vital of questions."

—*Psychology Today*

The

MIND

of

GOD

※ ※ ※ ※ ※ ※ ※ ※ ※ ※ ※ ※ ※ ※

Neuroscience, Faith, and a
Search for the Soul

Dr. Jay Lombard

FOREWORD BY PATRICK J. KENNEDY

HARMONY

BOOKS · NEW YORK

Originally published in hardcover in the United States
by Harmony Books, an imprint of the Crown Publishing Group,
a division of Penguin Random House LLC, New York, in 2017.

Library of Congress Cataloging-in-Publication Data is
available upon request.

ISBN 978-0-553-41869-9
Ebook ISBN 978-0-553-41868-2

Cover design by Jessie Sayward Bright
Cover image by Marco Brivio/Photographer's Choice/Getty Images
(The Sistine Chapel, Vatican City)

First Paperback Edition

147429898

To the memory of my parents,

Jeanette and Herman Lombard,

who saw the world as it was

and how it could be

The love of God for His world is revealed

through the depths of love that human beings

can feel for one another.

—MARTIN BUBER

Author's Note

In the interest of strict confidentiality, all reasonable measures have been undertaken to preserve anonymity, and thereby all patient names as well as certain details have been glossed for privacy.

This book is not intended as a substitute for the medical advice of physicians. Readers should regularly consult a medical professional in matters relating to their health, particularly with respect to any symptoms or conditions that might require diagnosis or medical attention.

One final note: Throughout the book, I refer to God by using the masculine pronouns *he* or *him,* despite the fact that our experience of God has both masculine and feminine attributes. I used these words for the sake of convention and sincerely hope that the reader will forgive me for this decision.

Contents

◇ ◇ ◇ ◇ ◇ ◇ ◇ ◇ ◇ ◇ ◇ ◇ ◇ ◇ ◇ ◇

The MIND *of* GOD

Foreword
Making the Invisible Visible

More than fifty years ago, my late uncle, President John F. Kennedy, made a public commitment to the American people to explore outer space within a decade. His administration also valued the "inner space" of the human mind, and he was a model for many future leaders, including me.

During the sixteen years I spent as the U.S. Representative from the First Congressional District of Rhode Island, I dedicated myself to achieving mental health parity, and since then, to support leading-edge research into the brain. The brain is an integral part of the body, to say the least, but it remains one of the least understood and most often ignored parts. We've got some brain-related challenges, both individual and societal. And just as we can find, explore, and treat physical illnesses such as cancer, we can begin to better handle the disorders and diseases of our most necessary organ, the one where, after all, the self and soul reside.

This book is about the exploration of our inner space. As a modern culture, when we explore our inner space, we can find vast expanses of the beautiful and uncharted unknown. But tragically, we often find incredible conflict within ourselves and between each other. We find stress. Anxiety. Anger, even baseless hatred. These are the most urgent diseases of our time. Just spend any random hour in front of a cable news channel and you'll see a worldwide epidemic of limited, lost, misguided, sick, or destructive minds bent on division. What can we do? We medicalize these problems, but the root cause is often existential, something deeper in origin. In this book, we want to explore what that means.

There is no health without *mental* well-being. Just as we have deeply delved into the fundamental origins of diseases like cancer, we need to explore, with similar drive and rigor, disorders of the brain. The magnitude and human cost of these conditions cannot be measured merely in economic terms; they touch upon the essence of our existence, the content of our lives. Untold numbers of people are suffering. Brain ailments are the number one cause of adult disability worldwide. Our children aren't immune, either, and early intervention is critical: More than six million U.S. children experience emotional or behavioral problems. Perhaps no class of citizen is more on the front lines of mental incapacity than those in the military. More than half of the most seriously wounded soldiers from our recent wars suffer from traumatic brain injury (TBI) and/or post-traumatic stress disorder (PTSD). And among *all* these people, 60 percent receive no treatment. Returning soldiers are all but abandoned in a byzantine system, which perhaps explains why 20 percent of all our suicides are veterans.

None of this bodes well. We have good cause to worry about our country's future. Children are increasingly anxious, stressed, and even violent, and among those who are treated, prescription drugs have become more and more permissive, even de rigueur. The first step toward addressing the root causes of these conditions is continued research, through projects such as the Brain Research through Advancing Innovative Neurotechnologies (BRAIN) Initiative, which is mapping and visualizing the circuitry of the brain, not only helping us to better understand complex human behavior, but also sparking profound new advancements in treating brain disorders. But we can't treat something we can hardly grasp—at least not optimally. That's the idea behind One Mind and other global initiatives for brain health that involve people working together to radically accelerate open science to benefit all people affected by brain illness and injury, as well as the associated intellectual and emotional disabilities. There is definitely hope, but we need to fully grasp the scope of the problem.

Across the world, neuroscientists in the Human Connectome Project (HCP) have spent the past five years mapping the brain, compiling an unparalleled storehouse of neural data on the astoundingly intricate web of connectivity that sits inside our skulls. This virtual brain mapping now gives researchers invaluable access into the living brain structure and function—and ubiquitous dysfunction. With at least a million-billion synapses, and thousands of miles of neural wiring—all of which guide our thoughts, create our feelings, retrieve our memories, and allow our consciousness to emerge—the brain is the most fascinating but still the least comprehensible of our organs.

Not for long. Through the kind of neuroscience Dr. Lombard is doing, we seek to bridge that gap; to identify, prevent, and treat major diseases; to prolong life; and, perhaps most important, to improve quality of life along the way. Finally, we seek to heal a world that's in considerable agony right now.

But the reason Dr. Lombard and I assign this vaunted status to neuroscience runs deeper, and at the same time it's simpler: Our mission is to collaborate, cooperate, and otherwise come together, not only to research causes and treatments of brain ailments but to prevent them from occurring in the first place. We achieve that latter goal by doing our part at every opportunity to make the experience of life for ourselves and others the least traumatic, painful, and meaningless. This, too, will be hard. Our suffering is great, and so widespread.

I know what this is like personally. I still need to manage my brain dysfunction. The most difficult obstacle I ever overcame was to turn my mental illnesses into a means by which to serve, to support others, and to work on becoming the best person I can be, for myself and my family. I have survived depression, bipolar disorder, alcoholism, and substance abuse. And through my own personal story, I've come to understand the necessity of ascension, both private and collective. First destigmatizing, then identifying and studying, and finally ameliorating brain disorders is a mission as imperative—or more so—than putting more humans on the moon or Mars.

To make major inroads toward a better world will require unprecedented cooperation on the vast frontier of brain research. This means bringing together every major organization related to brain research. We must pool our resources and form public and private partnerships that can unlock the mysterious vault of the brain once and for all, just as we've done with the heart, the pancreas, and every other vital organ.

Dr. Lombard and I are cautiously optimistic. Notwithstanding the enormous complexity of the human brain, we have never come closer than we are today to uncovering the astounding symmetry between our brain and the greater cosmos, the uncanny connections of the microscopic and macroscopic. This one organ, under God, with the liberty to fall prey to demons, or to follow a more healthful and productive path, is the single implement we will use to heal ourselves and others.

Coming together, as Dr. Lombard reminds us, means raising our hands high to confess unashamedly that we all have pain. We have unhelpful behavior problems. We have a tendency toward discord and war. We get our hearts broken when we lose the ones we love, as I did when my father, Ted Kennedy, died of brain cancer in 2009. This was the man who sat with me at my bedside through my frequent illnesses in youth. I thank God for that time.

We need to come together as families—it's not just our titles of familiarity that create lasting bonds.

We need to come together in our intimate relationships—we must find somebody, as I did in my wife, who shares our mission, who "gets" us and allows us to get them even if that means everyone must express vulnerability.

We need to come together as friends—I don't know what I would have done without the support from my Alcoholics Anonymous sponsor, a fellow lawmaker. It is possible to make friends with people even if we don't share the same faith or ideology. If one of our ideologies includes healing the world (or at least not making it worse for the next generation), then we can get along. If part of our faith is the faith we can place in each other to think of the "other" as "brother," then we are well on our way.

Finally, we need to come together as spiritual seekers: We need to deeply commit to entering into a respectful dialogue with faith-based leaders of every stripe, and to listen, learn, and provide the best evidence-based education about the brain. If we can bring together the adherents of science and the subscribers of faith for a substantive dialogue, *then* we'll be in business.

Just as spirituality is often necessary for addiction recovery in individuals, it will play a vital role in any recovery of our species from the ravages of brains gone bad, brains entrained toward hostility and nihilism.

How about if we reprogram our brains for love and service and other triumphs over misery? How transformational would this be—the ability to apprehend our *selves* at the most fundamental, existential, and theological level—for ourselves and the global community?

—Patrick J. Kennedy
Brigantine, New Jersey, July 2016

1

The Mind of God

If we discover a complete theory, it should in
time be understandable by everyone, not just by
a few scientists. Then we shall all, philosophers,
scientists and just ordinary people, be able to take
part in the discussion of the question of why it is
that we and the universe exist. If we find the answer
to that, it would be the ultimate triumph of human
reason—for then we should know the mind of God.

—STEPHEN HAWKING[1]

It was 1990 and I was in the beginning of my neurology
residency. Under the bright fluorescent lights of the pathology lab at the Long Island Jewish Medical Center, I
stared at the specimen that shimmered slightly on the
cutting board. I had examined several brains by then, but
this one looked different. At first I couldn't put my finger
on why.

The brains had been culled from the crop of patients
who'd died that week at our busy hospital. The neuro-
pathologist had laid out the brains meticulously, stems
and all, in front of us medical residents as she did weekly.
Her first task was to weigh this particular brain, and it

came in at a mere 1,005 grams, a good 10 percent smaller than normal. I peered down at the label on the cover of the clinical report—*Female, 5, Haemophilus meningitis.* My stomach sank.

It was a child's brain.

I wasn't prepared for that. This was the first pediatric brain I'd seen up close. What made my immediate task particularly difficult was that I knew whose brain it was. The brain belonged to a little girl named Sarah. I'd seen her only days earlier in the pediatric intensive care unit. A wave of sorrow washed over me, yet I felt another emotion, too—wonder. Sheer wonder. In front of me was the epicenter of Sarah's existence. Every thought this beautiful little girl had ever thought, every dream she'd ever dreamed, every hope for a bright and promising future had seemingly originated in this mass of protoplasm. I had to go forward. As is often necessary in my profession, I needed to check these emotions and steel myself for the business at hand. But my mind was on fire. Filled with questions that had no immediate answers.

When we examined this poor child's brain, we found widespread inflammation of the meninges, the protective membranes that cover the brain and spinal cord. From her cranial nerves to the base of her spinal cord, we discovered the remnants of the extreme inflammation that had occupied her skull. Sarah had died of textbook meningitis.

Sarah had had no prior medical history of meningitis, but that didn't surprise me. I knew the disease's onset can be almost apocalyptic. One day Sarah had been a perfectly healthy five-year-old playing at home with her family. The next day she'd experienced a headache that progressively grew worse. The headache was accompanied by fatigue, fever, nausea, and a general sense of unwellness. Her parents had given her Children's Tylenol initially and kept her home. (We've all felt these symptoms from time to time, and there didn't seem to be any reason for alarm.)

But when Sarah didn't get better after a day, her parents called their pediatrician, who recommended they take Sarah to the emergency room. No one knew at that point that Sarah was teetering at the door of death. Unfortunately, it proved only a short and tragic hop from the ER to the ICU, where Sarah quickly went downhill. Various tests were rapidly performed there in an attempt to find a solution. Then came a sudden attack. Within Sarah's medical records I read the frantic and heartbreaking attempts at resuscitation.

> Patient coded at 4:35 A.M. CPR performed for 45 mins. Epinephrine, dopamine, & external pacemaker. Unable to establish pulse or maintain BP. Patient declared at 5:18 A.M. Attending physician and family notified.

The words trailed off in my mind. This was finally how Sarah—her brain, at least—had come to the basement path lab. To me.

I went back to the task at hand. As I sliced into Sarah's brain, I marveled at how it was still a fleshy pink in parts—and that juxtaposition of pink, girlish life against the gray death of sliced brain matter struck me anew as an incomprehensible paradox. I stared at pieces of the brain, and the big questions of life flooded over me even in the midst of my dissection. Could the gross flesh of Sarah's brain really have possessed the whole elaborate fabric of Sarah's very being—her emotions, memories, hopes, and all the intangibles of her existence? To think that this little girl ate, slept, dreamed, imagined, laughed, told jokes, hugged her parents, and formulated her entire conception of the world around her owing to the power of this mere blob of meat. Surely residing within or perhaps beyond the organ of the brain was something greater, an immaterial primary essence of a person that still survived.

The light above me flickered in the path lab, threatening to darken the table, the brain in front of me. I considered how billions of people cling to the idea of some kind of afterlife.[2] Perhaps they are simply deluded or in denial. But what if they aren't? What if an afterlife is actually plausible? What if we could create a case for it using the rigors of science?

My questions didn't stop there.

Here I was, a young physician beginning his life's work in a world of objectivity, only evaluating things I could quantify, and I was faced with life's biggest queries: Was there some other truth I needed to find for myself? Surely there had to be something greater to the end of a human being's life than what Hamlet would call "bestial oblivion." But what was this "something greater"? And could I reconcile it with what I knew of science?

What Made Sarah *Sarah*?

Fast-forward nearly three decades since that day in the path lab. I find myself still a neurologist steeped in neuroscience, still actively engaged in a concrete world of measurements and data. Yet those questions about something existing beyond flesh and form still persist.

Simply put, neuroscience is the study of the brain—it is the portal through which we can discover, in the brain's formerly fathomless code, the true nature of our being. Yet we are much more than mere intellect, more than human computers who make decisions. The more I've discovered about the living brain, the more secrets I've unearthed about the nature of human beings, the universe, the purpose of our lives, and the possible existence of something beyond all of this. For me, neuroscience isn't just the study of the brain; it is a tool by which we can make the invisible visible. As Antoine de Saint-Exupéry wrote in his masterpiece *The Little Prince*: "What is essential is invisible to the eye."[3] No truer words have ever been written. Through an exploration of our brains, we can discover more about the nature of faith, belief, and hope. Thus, there is logical and rational evidence that can help us answer the big questions of life.

In the pages to come, I want to take you on a cerebral journey—both literal and figurative—into the depths of the mind, to explore the most important questions about life and how neuroscience may help answer them:

1. Is there a God?
2. Do humans have souls?

3. Are we special? (Meaning, are humans any
   different from other animals?)
4. Do we have free will—or is all of life
   predetermined for us?
5. What's the meaning of life, and is there any
   higher purpose for our existence?
6. Since evil exists so prevalently in the world,
   can there be such a thing as a good God?
7. Is there life after death?

These are questions that have long challenged me to search for answers. Certainly, many contemporary neuroscientists will disagree with my conclusions; they'll argue, for instance, that a person's consciousness—the implicit and yet ubiquitous "awareness of being"—comes down to neurological switches that are flipped on or off, as in an electrical engineering circuit. For many of these intelligent men and women, whatever happens in our mental state is purely a function of physical (neuronal, biological, or electrochemical) reactions. The general belief among scientists, doctors, and other medical professionals is that we are purely material, with no soul to speak of beyond the elementary particles of chemistry and biology. The majority of neuroscientists assert that there is no physical evidence of an existence beyond the flesh, that any person's subjective sense of self as a soul or immaterial being is an illusion, a simulation designed to hide the actual backroom computations and neuronal workings of our brains.

Yet when discussing these topics with my colleagues, many of the most humble recognize there are gaps in conventional scientific thinking, spaces yet to be filled in the scientific world—and what can fill those gaps leads back to the reality of the intangible. These gaps are referred to as the "hard problem of consciousness,"[4] a term coined by the eminent philosopher of mind David Chalmers to describe the vast chasm between the physical and the phenomenological, between the tangible senses and the suprasensory experience we have of them.

Let's look at this a little more closely. While our brains are indeed biological, the experience generated by our brains—our thoughts, feelings, and beliefs—is beyond the observable and measurable. Our subjective perceptions, the qualitative and sentinel aspects of our being, cannot be merely by-products of neurochemical processes. "Even if it were able to map out the precise pattern of brain waves that underlie our subjective states, that mapping would only explain the physical correlate of experience, but it wouldn't be them. A person's experiences are as irreducibly real as her brain waves are, and different from them."[5] Or to paint another picture, if we are all fish in the fishbowl, how do we observe ourselves outside the fishbowl?

Consciousness is not something to be feared or dismissed. The Latin roots of the word *consciousness* mean "to know together" and hence this distinctly human attribute is a gift that provides us with the capacity to jointly inquire about the meaning of our existence. If we were able to unwrap this gift, we would see that the perceived gaps between flesh and spirit, mind and brain, hide a deeper, intrinsic, and fascinating reality. This reality is an experience of something ineffable—a mind, soul, spirit, or even energy—that which is both irreducibly complex and fundamental to our being.

Is it what made Sarah *Sarah*? What makes me *me*? What makes you *you*?

Faith in Science

Now, I am first and foremost a neurologist who is deeply vested in empirical data. I hold great faith in science. I, and millions of other physicians and scientists, have staked our professional careers on the integrity of the scientific method, which has proved to be extremely useful to help predict and manipulate natural, chemical, and biological phenomena. By no means do we want to ignore science in this process of discovery. To the contrary, we want to use science as a stepping-stone to learn and know and grasp all we can about who we are. I realize the very idea that truth can be found both "within" and "beyond" science can be difficult, if not impossible, for some people to grasp, particularly for us scientists—remember all those fish in the fishbowl. But with science, faith, and a little bit of reasoning, we can press our faces upon the glass and see that there is something beyond, and it's extraordinary.

Science Is the Only Truth?

As I've alluded to earlier, there are many who believe that science is the only truth. Yet truth can be found both "within" science and "beyond" science. Truth can be found in philosophy, literature, art, music, and history.

Truth can be found in the stuff of life that's beyond what is fully measurable or seen. The famous physicist Max Planck expressed a similar sentiment: "Science cannot solve the ultimate mystery of nature. And that is because, in the last analysis, we ourselves are part of nature and therefore part of the mystery that we are trying to solve."[6]

In other words, fish in a fishbowl.

Some people will label these further forms of truth the same as faith, but I'm not completely comfortable with that term either. *Faith* is such a loaded word, so personal and charged, so full of various connotations. People talk about taking a blind leap of faith, or stepping out on faith into the great unknown. But that kind of faith is far too limited. The kind of faith I'm talking about in this book doesn't ask us to suspend facts. By contrast, it calls us to examine the facts and then build upon them. The faith I'm talking about (if we even want to call it faith, but for argument's sake we will) is a faith informed by science, measurability, and logic, not by blindness. This faith asks: What if there were actually very good reasons to believe in the intangible? What is faith? Faith means accepting that there is a greater reality beyond our senses and our intellect. We can use our intellect to instruct us on how we should live our lives, but faith teaches us about the meaning and purpose of our existence. What if there are very good reasons to hold to and live by the veracity of that which cannot be seen?

As a scientist and as a human being who has grappled with the meaning of faith in my own life and with many patients struggling to comprehend whether life has some deeper significance beyond the here and now, I believe

it is possible to develop an appreciation of both the biological and the transcendent, and to explore how each
can inform the other in positive ways. There must be a
balance.

We must have science.

And we must have faith.

Caveats and Definitions

Right up front, please know I am not calling anybody
toward a particular form of religion or belief system.
Nor am I calling people toward the specific definitions
or descriptions of God found in any historic faith-based
book. I am not trying to convert anybody, and this is not
a book that trumpets any specific religion. It is a book
about large faith, this hybrid sort of Mind-of-God-styled
scientific faith whereby we seek to understand the meaning of our lives through the portal of neuroscience and
what we have discovered about the operations of our own
brains. My main reason for writing this book is that I believe humans as a species will be better off with this sort
of faith in mind, a faith invigorated and enlightened by
science rather than being at odds with it.

There's a greater narrative here that offers to bring together the many fragments and threads of our biological
and immaterial existence.

As I've already mentioned in part, I'm pretty sure we
will never fully intellectually or philosophically or psychologically or neurologically decode the essence of God,
nor should we waste our time arguing about this or that.
We should, on the other hand, be concerned with how

we live our lives and offer service, kindness, love, healing, and forgiveness to others. Our actions through faith will always speak louder than the ideological words we use to describe them.

So the premise of this book is to help us understand the brain better. By doing so, we can understand the mind better (brain and mind are not the same—more on this later). And by doing this we can muse on and have a conversation about what could lie beyond our mere biology. My hope is to help us appreciate the intangible, which in turn should help us grow closer to each other, respect each other, and ultimately heal each other through embracing faith in our lives.

What I am arguing for is an active faith—not a passive belief in superstition or surface ritual. Active faith entails seeing beyond our differences and embracing the truth of connectivity: that our actions impact others. If you were thirsty on a scorching summer's day and I brought you a cup of cool water, then I would be practicing my religion, and in that sense I am fulfilling the will of the intangible. I am fulfilling the will of God. By offering you the cup of cool water, I am engaging in empathy and goodwill toward others. I am connecting to another person in a way that goes beyond what many believe is our strict evolutionary nature to dominate, reproduce, and survive as one of the fittest of our species.

The big takeaway of this book is highly practical: a change of heart. Empathy. Altruism. Compassion for others. The more we can participate in a larger active faith, a belief and truth in something beyond the material, the more understanding and considerate of other people we can be. This sort of faith can unify us as people, not tear us apart. It's in our relationships with each other that we find our collective truth and purpose. I will come back to this point again and again in this book: Faith must lead to compassion. It must! Or else what good is faith? We cannot grasp the infinite Mind of God except through our capacity to deeply love and understand each other.

Faith and science can go hand in hand, and I don't see an unbridgeable gap between the two worlds of God and evolution. Of course I hold to evolution. Everything is evolving and adapting, but I would hold that our universe was created with intent, an "intrinsic potentiality."[7] The smashing of sheer randomness ex nihilo cannot create the complex order we find in the universe today, so it's highly plausible that our universe was originally created by some sort of first cause, be it personal or otherwise. Within the big bang, there must have been some set of cosmological blueprints that instructed galaxies and atoms to behave as they do. How did those blueprints spring into being except by some sort of intent? Does it really make sense to suppose that the world in itself, without the presence of God, should be doing something we could sensibly call "aiming at" some states of affairs rather than others?[8] Goals imply agency, and agency is evidence of mind. The eighteenth-century German philosopher Immanuel Kant, one of the greatest thinkers of all time, set a similar thinking in motion. He wrote:

> In the world we everywhere find clear signs of an order in accordance with a determinate purpose.... The diverse things could not of

themselves have co-operated ... to the ... determinate final purposes had they not been chosen and designed for these purposes by an ordering rational principle in conformity with underlying ideas.... There exists ... a sublime and wise cause ... which must be the cause of the world not merely as a blindly working all-powerful nature, ... but as intelligence, through *freedom.*[9]

The Origins of Disease

In my neurology training, we are taught that when diagnosing a brain disorder, we rely on reviewing all the signs and symptoms a patient presents, as these will help us discover the location of the disease. By localizing where in the brain a particular disease resides, we can diagnose it better and develop more effective strategies to treat it or prevent its recurrence. In other words, find it, name it, fix it.

But how do we find the location of a disease that is intangible—and by that I mean cruelty, pain, suffering—the roots of which are existential? We can measure all the salt content of our tears with the most sensitive instruments yet never truly understand the meaning of sadness. How does one quantify hope or faith or love or joy? How do we find the origins of hopelessness, nihilism, or destructive beliefs to weed them out? The need for faith is so deeply embedded in our biology that even if we don't identify it, its absence will let itself be known regardless. I learned this lesson early on in my training, but it wasn't taught in any textbook.

Imagine, again, Long Island Jewish Hospital. I'm an intern. It's 1989. I was assigned to care for a woman with advanced metastasized ovarian cancer. Eva was bedridden, jaundiced, and suffering from ascites, an accumulation of fluid that produces abdominal swelling. At this late stage, there was really very little we could do for her other than treat associated infections, try to make her as comfortable as possible, and reduce her pain. It was really only a matter of time.

Every day, first thing in the morning, I went to her room and spoke with her, checking her vitals, doing her blood work, holding her hand, always trying to encourage her and give her hope. I wrote brief progress notes about her status. We talked about the weather, the nurses, the news. She always smiled whenever I came in, and she shared a corny joke or two with me from a rich storehouse of humor. I asked her about her family, the various faces that beamed from photo frames on the windowsill, and she talked freely about the people in her life who meant the most to her. If a wave of pain swept over her, I could always dam it by asking about her grandkids. One played hockey. Another was a budding baker. A smile would come over Eva's face as she lovingly described each one.

For twenty-five days I did this—and for those twenty-five days my patient lived. But two things began to gnaw at me. First, this type of medical work typically is not considered high priority for a first-year resident. We doctors are all a bit arrogant at this early place in our careers (I know I was), and I got to thinking that all this talking work could be better performed by a social worker. Surely treating the "real" patients who could be served medically was a better use of my time. Second, I felt that unspoken and horrible pressure that every intern feels to get a patient off his rotation list. When you have fifteen to twenty patients in your care, the more patients you have, the more difficult your workload is. It's simple math. You are under pressure to lighten your load by discharging your patients as soon as possible or by transferring them to a different service in the hospital. It's not a good goal, I know. But it's a pressure we all felt.

On the twenty-fifth day, I told my chief resident there was unfortunately nothing else we could do for Eva. It was the stark truth. The next stop for Eva was either being transferred to hospice care or going home to die. The chief resident called in Eva's husband and discussed options with him. I remember watching the husband from behind a privacy curtain and hearing his reaction. He was very upset. No one wants to receive news such as that. I went home that evening, then came back the next morning. The twenty-sixth day.

And overnight Eva had died.

Technically, I had done nothing wrong. I knew that. But the timing of the decision to discontinue care and Eva's passing felt too coincidental to me. I felt guilty. Every day for those twenty-five days when I had visited her, I had given her hope. Then, with my decision to transfer her out of my care, her hope had vanished. When we told her husband there was nothing more we could do for her, I had inadvertently pulled the plug on my patient. Eva wouldn't have lived much longer either way. But that hope could have bought her a few more days at least. Perhaps even a few weeks.

Hope: invisible, intangible, sometimes impractical.

It struck me then, for the first time in my medical career, how important the power of faith is to sustaining life—the very potent but invisible forces we cannot measure or quantify, yet which are essential to the very core of our being.

As a young doctor, I was on my way to making my living from immersing myself in a world of facts—that which could be seen under a microscope, viewed on a chart or a test, or prescribed and delivered in a bottle. Yet when it came to Eva, I had greatly undervalued and minimized the nonclinical side of medical care. My visits to her, my asking about her children and grandchildren, my sharing in small jokes with her—these "housekeeping" matters carried much more importance than I'd thought. In Eva's case, the tool of "hope"—the expectation that there would be a tomorrow and a next day—became an influencer on how she lived and how she died.

Decades have passed, and what I learned from Eva's case continues to affect me. I've treated thousands of patients over the course of my career, and I continue to be immersed in a world of facts, yet it's the things that can't be measured that continue to hold great fascination for me. Too often we doctors believe that if we can't measure something, then it's not real or doesn't exist. But before Eva's passing, I'd encountered something buoyant. Something real and powerful.

And it could not be measured.

I'm desperately glad for the full arsenal of quantifiable medical procedures at my disposal to use for my patients' benefit. Just think, where would we be today without the MRI machine? Or the Doppler ultrasound (used to determine the risk of stroke)? Or the electroencephalogram (a test that measures and records the electrical activity of the brain)? I love facts and I love science and I love the amazing advances that have been made in medical care over the years. Yet today, as I approach my late fifties, I am more convinced than ever that essential things exist even though they are invisible to the human eye, and this awareness compels us to inquire into what Thomas Moore aptly described as "the paradoxical mysteries that blend light and darkness into the grandeur of what human life and culture can be."[10] After all, some of the greatest masterpieces of art are those that show an understanding of this relationship between light and shadow.

Sacred Mind

My hope for this book is on an individual level for each reader, regardless of any particular faith. It's for *you*. I pray that you will not lose faith because of science. Rather, I entreat you to regain a larger faith through science. Faith and science are not incompatible. They're integrated systems, not separate.[11] Ultimately, I hope that you will find a deep sense of purpose in your life and relationships. If you want more meaning in life, an understanding of life that goes beyond the here and now, or if you want a deeper relationship with God and a closer relationship to others, then my invitation to you is to start probing the Mind of God through the window that is your own mind, and thereby discover a mind that is fruitful and constantly embracing and ultimately creates and doesn't destroy. When you think with this position of openness, you can radically and positively change your life.

Neuroscience offers us a window into the material basis of our immaterial existence. As the Lubavitcher Rebbe once wrote, "reality is not a product of our mind, but ... *the mind is a product of this reality.* Reason may lead us to the door of this reality, but we need different tools to enter."[12] It's with this twinned medium—the human brain and mind—that we are imbued with the capacity to inquire about the deepest existential issues we humans have ever asked and sought to answer—questions about our purpose, meaning, and agency. This endeavor is as ageless as the history of our humanity, and it is nothing less than the search for the human soul. Through hacking the hidden code of our brain's software, we are likely to encounter the wizard behind the curtain. And if we by chance happen to meet this wizard (as we'll explore in the next chapter), what would he reveal to us?

Quite literally, it's the ability to find our way home again.

Does God Exist?

Where is the knowledge we have lost in
information?

—T. S. ELIOT[1]

Does God exist?

It's a question that goes to the core of our existence, an inquiry that billions of us ask at least one time in our lives. Yet before we tackle that impregnable question, I would like to share a case of the most unusual and anomalous pregnancy. The story about this expectant parent is perhaps the most curious anecdote of my entire career.

He was gleeful. Mysteriously gleeful. Far too gleeful given the condition he was in.

John was a scruffy late-twenties man, restrained to a chair in his hospital room and wearing nothing but an adult diaper. His belly protruded significantly, and when I walked into his room he patted his belly lovingly with his one free hand and gave me a big grin.

"Hello, John," I said. "I'm chief of neurology here at Bronx-Lebanon Hospital, and I'd like to ask you a few questions."

"Okay," he said, smiling. "But you gotta hurry." He patted his belly again.

"Why? You going somewhere?"

"Delivery room! Can't you see, Doc? I'm having a baby!"

"A baby? Well, that's quite remarkable."

"I know! I don't really seem like the parenting type, do I? But I really think I'm ready now. When I was younger, I ran around a lot. But it's finally time to settle down with this little one." He stroked the side of his belly as an expectant mother might do and gazed lovingly at his protrusion.

I had to admit his face genuinely glowed.

I knew John had been diagnosed with multiple sclerosis (MS). The fact that he appeared unconcerned about his arm and leg restraints was scientifically explainable. A disorder called la belle indifférence is sometimes associated with the disease. La belle patients typically display an inappropriate lack of concern or emotion in proportion to the gravity of their symptoms, or in response to others' concern for their disability. But the possibility of la belle disorder didn't explain John's pregnancy fixation. If I was to truly help this patient, I needed to understand how he could believe that a male of the human species could find himself with child, something that even a person with dementia would be unlikely to believe. Did John have a gender identity disorder? Had he completely dissociated from reality? I decided to probe his psyche more deeply.

"John, you realize what you're saying—"

He interrupted, "If you're asking whether I'm nuts, I'm

one step ahead of you, Doc. You just have to chalk this up to a miracle. I don't understand how I became pregnant myself. It just happened. The important thing now is that we all make ourselves ready for it. For my part, I stopped smoking reefer and drinking beer. I'm taking care of myself for the first time. Gotta get clean so my baby is healthy." He rubbed his abdomen again, then scratched his stubble. "Oh, I can feel it kicking!" he added, grabbing just above his navel.

I suspected what he was feeling was more intestinal than gestational, but I kept my mouth shut. My mind continued to work through possible diagnoses. A woman can sometimes develop pseudocyesis, popularly called "phantom" pregnancy or "hysterical" pregnancy, a condition in which she will develop all or some of the psychological and physical symptoms of pregnancy. It happens in about 3 out of every 20,000 live births. Men can develop a related syndrome called couvade syndrome or "sympathetic" pregnancy, where the male partner of a pregnant woman experiences the somatic symptoms of his partner, such as backache, nausea, and even weight gain. But John was single. And when I questioned him further, he indicated that he didn't have a girlfriend or any other female in his life who was pregnant by his doing.

We talked for a while more; then I left for the lab. Tests soon revealed there was no miracle baby inside the man's belly. Instead, a much more ominous affliction had caused the distension of John's abdomen: liver failure causing ascites. But the liver failure was not consistent with the original diagnosis of MS. Something else was going on inside John—but what?

Our team reviewed John's MRI studies, and while he showed some symptoms of MS, the pattern wasn't consistent overall. His initial diagnosis was incorrect. We sent off more lab work, including an analysis of very long-chain fatty acids (VLCFAs), and developed a hunch that John had a different, much rarer disease, one that can mimic MS. Lab reports later confirmed our hunch.

John had a rare disorder called adrenoleukodystrophy (ALD), the disease featured in the 1992 film *Lorenzo's Oil*. It's a genetic disease characterized by the loss of myelin, a protective sheath surrounding nerve cells in the brain, and the progressive deterioration of the adrenal gland that soon follows. Though this result explained what we saw on the MRI report (and what had led to his stomach protrusion), it still didn't explain why John believed he was pregnant.

In a female experiencing pseudocyesis, a physical cause is to blame. There is a chain reaction that takes place. The brain is so intensely connected to the body that if a woman greatly desires a baby, she might begin to feel subtle physical signs that she has in fact conceived— slight weight gain, swollen breasts, and even the sensation of fetal movement. In turn, the brain can translate those incoming messages from the sympathetic nervous system as the real deal. The brain will then release various pregnancy hormones, thus exacerbating the initial "symptoms" and causing real signs of pregnancy.

But in John's case, no one had taken his delusions seriously enough to explore their possible meaning. The simplest explanation was schizophrenia, and we were otherwise too busy exploring the organic causes of his distended belly. We finally traced the (strictly) physical causes that had led to his pregnancy fixation. With the onset of ALD, John had suffered an accumulation of VL-CFAs in his brain's white matter. As a result, his brain was now damaged—but only partially. Reports show that other patients with adult-onset ALD had also experienced psychiatric symptoms due to the disease, although in John's case, for whatever reason, not all of his brain was damaged yet. The disease had predominantly damaged his brain's right hemisphere, the side of the brain that alters a person's experience of the world, starting with his understanding of his condition.

To explain—in each hemisphere of our brains, we attend to and construct a representation of reality that is complementary to the other side's reality, but dichotomous. The differences between our left brain and right brain are the distinctions between facts and their meanings. The left brain's primary role is to objectify events, and the left hemisphere's function is generally associated with language, calculations, and logic. Usually when we speak of accumulating or retrieving facts, we are referring to processes occurring predominantly in the left half of the brain.

By contrast, right-brain functions fit more in the realm of understanding the interpretative and sentinel value of our experiences. As the right brain seeks meaning, its drive is to discover connections, networks, and relationships.[2] A person's right brain functions like a GPS does for a car. It lets a person "see" and "understand" his or her relative location in time and space. It takes an experience and gives it context, meaning.

But since John's right hemisphere was damaged, his left hemisphere had taken over and filled in the gaps by feeding him the most logical explanation it could dream up for what was happening to him. John saw himself in the hospital surrounded by machines. He saw his belly protruding significantly. His GPS was damaged, so his left brain took over and concocted the most plausible explanation.

This is what John's "disabled GPS" told him:

A: Pregnant people's bellies protrude.
B: Pregnant people go to the hospital.
C: I am in the hospital and my belly protrudes.
   Therefore I am pregnant.

It was false logic, of course. But since John's brain now had an explanation, it convinced itself it was right and stopped asking questions.

Undoubtedly his brain's false conclusion was also exacerbated by social isolation. The entire time he was in the hospital, John hadn't had a single visitor other than medical staff. He needed a close friend or relative whom he believed he could trust explicitly. He needed the strong relational ties that offer a person a greater purpose to go on living. To compensate, his brain created an alternative reality in which he was not dying alone in the hospital.

John's is a troublesome and poignant story of a man dying alone, and I realize it comes in a chapter entitled "Does God Exist?" This begs the question, of course—what's the connection? What does John's story have to do with the larger questions raised in this book?

Or let's put this in personal terms: What does John's story have to do with God being real to you?

A Blindness of the Soul

Gabriel Anton, a renowned Austrian neurologist, wrote about a blind woman who did not recognize her condition. Although she had lost her vision, she did not accept that she could not see. Let that last sentence sink in for a moment, for in addition to holding great literal ramifications for the woman, it holds profound metaphoric ramifications for you and me.

The physicians caring for this woman were utterly confounded by how it could be that she was unable to notice her massive and complete loss of vision. How can a person be blind and not realize a condition of darkness? Anton wrote in a medical journal, "She assured in a calm and trustful way that she saw objects that were shown to her, while the everyday examination proved the opposite."[3]

Anton called his patient's condition "soul blindness," which was later changed to the term we use for this condition today—anosognosia ("lack of knowledge"). The condition typically stems from an injury to the right side of the brain and is commonly associated with a lack of awareness of one's deficit and the ensuing explanations, often of a confabulatory nature, surrounding it.

The brain is a marvelous but still deeply mysterious organ. Strangely enough, we know what the brain does mostly from studying what happens to it when it gets damaged. In a similar condition, when the right brain becomes damaged in its visual processing centers, it can result in "cortical blindness," in which a person's eyes still function normally, but the person is nevertheless unable to see. Light hits the optic nerves. Raw data is transmitted to the brain, yet the affected person has no explicit awareness of the light she is seeing. Because of the injury, the brain is unable to consciously access or articulate the information being sent to it and will convince itself of some alternative, prefabricated reality in its stead.

One of my patients had an extensive right hemispheric brain injury resulting from a stroke. I witnessed how she would apply makeup to only half her face as if a big vertical line had been drawn from her head to her toes. She had lost all sense of half of her body, even denying that her paralyzed arm belonged to her! How could a person walk around believing she was only half a person?

So let's offer this big question: What if greater realities exist in our universe (or outside our universe) than we can actually understand, comprehend, or measure with only one of our brain's two hemispheres? In other words, what if God exists but our brains cannot comprehend the sheer magnitude of light, beyond the spectrum of any perceived reality?

My invitation to you in this chapter is to take a closer look at the design of the right brain and its unique capacities. We must probe the hidden treasures and unknown potential the right brain offers if we are to understand our unique place in the universe, how we perceive it as such, and our potential understanding of God. John had facts alone, and he was seeing some realities. He was seeing the hospital, and he was seeing his extended belly. But he wasn't seeing the whole reality. Thus, he was "blind" to the reality of his life.

When it comes to our seeing the greater truths of the existence of God, I pray a similar thing will not happen to us.

How Can We Know the Unknowable?

Recent brain imaging studies demonstrate that our brains are hardwired for faith. Brain imaging has been done on Buddhist monks when they meditate, on Catholic nuns praying, on mystics doing contemplative practices, and more.[4] These studies frequently exhibit discrete and preferential right-sided brain activity that can be objectively measured.

Similarly, clinical imaging studies have shown that the area of the brain activated during subjective religious experiences is the same part of the brain that allows us to be aware of others—to be empathetic, emotional, altruistic, thoughtful, and aware of ourselves. All of these functions take place in certain regions of the right hemisphere.[5] It's almost as if nature has reserved an anatomical sanctuary where our brains can transcend the everyday elements of our existence. We meditate and pray with the same side of our brain that we use when we're kind to other people. A clue, perhaps?

Our left brain will insist that a belief in God is not enough to know of his existence. There must be more proof. But what the left brain cannot deny are right-brain processes—that belief, faith, love, hope, and an ability to see the "big picture of life" are biologically programmed in our right brains.[6]

Think of it this way. Perhaps we simply imagined God into being. Humanity wanted something greater than itself, so we made up this concept of divinity. From a sheer biological perspective, we must ask how our brains had the tools to do this imagining in the first place. Inanimate objects such as rocks and trees don't have a recognizable concept of God. They have no empathy. No relationships. No awareness of others. Only humans can have a belief in God. Why? How did our brains ever have the tools to develop faith in the first place?

The answer to that may well lie in an epistemological exercise I call the neuro-anthropic principle, a hybrid of other existing principles. (*Neuro* means that which can be detected in the brain or nervous system, and *anthropic* relates to human beings or their span of existence.) This "argument from design" is by no means new. Also known as the teleological argument, it spans the time from the metaphysical philosophies of Plato, Aristotle, and the Western Enlightenment, when anyone who meditated long enough on the elegance and cosmic majesty of the universe might be propelled toward conviction about the existence of some Mastermind behind the curtain.

The more recent anthropic principle, as promulgated by Australian theoretical physicist Brandon Carter and American Nobel laureate physicist Steven Weinberg, supposes that the world is friendly to life and exquisitely tuned to accommodate us; that among the living things it supports, only one of its chosen recipients can ultimately conceive of, measure, contemplate, and settle on a belief about these circumstances; and that this medium is our human mind.[7]

Simply put, we live on a custom-made planet that is friendly to life. It's sublimely ordered to afford the improbability of life possible. Goldilocks couldn't be any happier with Earth. If Earth were only a little too hot or a little too cold (or anything else), then we literally wouldn't be here. But, like Baby Bear's bowl of porridge, Earth is "just right." That means that life requires a multifarious matrix of exactly occurring, precisely timed, epic and epochal, suspiciously fortuitous circumstances, and it's the human brain most apt to recognize the miracle of all this.

A key question is, Must all of this have happened by design? Could and would these conditions have developed without us? The evidence suggests the universe needs us as much as we need it—we co-create our existence through our conscious perception of reality. Stanford University physicist Andrei Linde concluded that the universe was designed for us to recognize it: "The universe and the observer exist as a pair.... I do not know any sense in which I could claim that the universe is here in the absence of observers."[8] Certainly it's a bit of the lone-tree-falling-in-the-forest question (can anybody hear it?), yet what Linde means is that the twinned mediums of universe and mind suggest no coincidence at all. There is a sense, yes, in which our minds are required to bring the universe into existence.

What does it mean when we say things such as that we "co-create our existence"? Or that we "determine our own reality"? It means, simply, that things exist because the mind makes them so. Our perceptions create our circumstances. Our thoughts create our realities. But with a few caveats. It's popular today—although erroneous—to conclude that since we determine our own reality, we therefore determine God. We create God, so we are the same as God. Many faith traditions find this thinking heretical. In truth, we are not the ultimate determiners of reality. We are not the Creator. We are the creation. So in this context we mean "co-creation" in this sense: A child will say, "I don't like vegetables." She is therefore determining her own reality. Has she tried vegetables? No. Well, how does she know she doesn't like vegetables? Because she has decided she doesn't. Her mind has developed her own reality. Her mind has created her "universe."

Cosmologists seek the origin of the universe. Archaeologists seek the origin of civilizations. Evolutionary biologists seek the origin of mind. In each of these quests, experts look to uncover the vestiges or remnants of what has passed, to trace not only the physical evidence but the essential conditions that aggregated to shape who we are today, including how our brains have evolved over time.

The resulting question is this: What evolutionary significance drove nature to shape our brains in such a unique and specialized manner that we developed a concept of God (as well as our ability to deny his existence)? What purpose does belief have, and do we understand enough about our brains today to objectively inquire about not only the facts of human experience but also their meaning? As the right hemisphere of our brains is designed for relationship and connectivity, are there some buried secrets and memories of a more elemental and unmitigated experience of God contained within us but with a fading accessibility? Can this fundamental remembrance lead to an unexpected spiritual awakening?

All acts of creation—the conception of a child, the formation of a persona, the birth of a star—require a separation of space, a division of time, and an individuation between self and other. There is inherent trauma in every delivery, some element of dissociation intrinsic to every division. This is no less true in the evolution of our humanity and, of course, human consciousness, our "expulsion from Eden." Creation entails an array of circumstances in which what appears indivisible becomes divided and partitioned. So, too, our brains allow us to perceive and therefore believe that our existence is also one of division—separate and apart, autonomous, alone.

Separation Anxiety

It is within the mind's capacity to recognize its ultimate origin and purpose, the recognition of an intrinsic and implicit connection to that which is beyond itself—made possible by an essential vestige of our own brains—even if we experience separateness. We are born with the capacity to believe even if we choose not to.

Some will argue that this is fallacious, because our brains have the capacity to imagine all sorts of myths and implausible situations. We can imagine dragons, for instance, but dragons don't exist today and never have existed. Why couldn't it be possible, therefore, that we have simply imagined God? How does our knowledge of our brains and the evolution of human consciousness postulate the existence of God?

Our brains are miraculous in and of themselves. Not only are they specifically designed to contain facts; they also have the capacity to interpret them. These interpretations will always remain subjective but no less valuable to explore than our quest for objectivity. The brain's "receptors" for these experiences are the root cause for our myths, our stories, and our beliefs. These uniquely human attributes—the inherent capacity to encounter the transcendent and immaterial—bridge the gap between faith and reason, between subject and object, between self and other.

The hybrid theory, the neuro-anthropic principle, posits that our minds were designed in such a way as to acknowledge our unique purpose in the universe, to recognize our highest connection to God and to each other. We are born with the capacity to make the invisible visible, to have faith in the intangible. Saint Augustine writes that the human mind is worth more than the whole of inanimate creation.[9] The mind can observe the latter, explore it, measure it, wonder at it—but the inanimate universe cannot do the same.[10] Our brains reflect a universe so designed to reflect it.

This immense worth of the human mind means that we can see and feel, in real and tangible ways, that we are the cause of each other's suffering. Because we feel, we know others feel. This lets us understand that psychic pain is as biologically real as cancer. I'm speaking here of that portion of our brain that inherently possesses faith in something beyond itself, a "supra-sensory" and higher form of knowing than our perceptual biases of division lead us to believe. When we nurture that aspect of our being, we are able to personally actualize our higher purpose, which is to become agents of mercy and compassion, to connect to that which is beyond our mere physical existence.

Let's look at an example from the realm of neuroscience that relates to this deeper sense of knowing. A developmental psychologist asks the question: When does a father become a parent? The simple scientific answer is: at the biological moment of conception—that is, when a sperm fertilizes an egg. Yet as a neuroscientist, I can see that conception alone doesn't make a parent in the fullest sense of the word. Not at least as far as the brain is concerned.

Rather, a man becomes a father in the fullest definition of fatherhood when he holds his child in his arms. The brain does something different then—and we can see this difference in brain imaging tests. Wrapped in the man's embrace, the child knows he has someone who will take responsibility for his well-being, and the man's brain "co-creates" his role. He is not just a biological procreator; he is a parent. A new, detectable, and powerful neural attachment to another human being occurs. Neurologically, this knowledge takes a form that's quite different from, say, grasping the day's weather. The knowledge a man has of being a father after he's held his offspring is substantively different from the knowledge he had before a child was present to hold. And how these two kinds of knowledge differ is a fascinating question that neuroscientists have spent a good deal of time learning more about—where there is one aspect of knowing that focuses on isolated facts and a whole other type of understanding that embraces the interrelationship, meaning, and implicit value between things.

It turns out that the different sides of our brains possess different capacities for knowing. The mere knowledge that the process of mixing a male sperm and a female egg will equal a baby is very different from the deeper knowledge and feeling that "this child in my arms is mine." The former is knowledge typically associated with left-hemisphere functions—the basic facts of the matter. The latter is knowledge typically associated with processes in the right hemisphere—the value, meaning, and purpose embedded in these facts. Take a moment to close your eyes and imagine someone you love deeply. You just had a glimpse of your right brain in action.

The uniquely specialized capacities of the right hemisphere such as recognizing a face, comprehending imagery, recognizing context, and interpreting tone help us maintain and sustain relationships and develop beliefs. The right hemisphere's functions are essential if we are to put together the aggregate and fill in the gaps to see the whole.

With the parent-child bond, something neurological is happening to the baby, too. When a baby gazes into his father's eyes, the right side of the baby's brain is imprinting crucial social information about what the man looks like and whether the baby can trust him. When the man gazes back, primarily the right side of his brain processes the gaze, which calms the man and stimulates his nurturing behavior. In an ideal attachment situation, both parent and child utilize right-brain capacities to form a social bond and begin a lifelong relationship of understanding and trust.

The functions of the left and right brain are not always this distinct. Rather, communication between the two hemispheres occurs through the corpus callosum, the bundle of neural fibers that facilitate the connection between the two separate regions of the brain. For example, the left side of the baby's brain might pick out the sounds of speech and how to put those sounds together to form words. Meanwhile the right side homes in on the intensity and stress of the language's rhythms. With our left side and right side working together, we grasp not just the content of what's being said but also the speaker's underlying emotional meaning and feeling.

The code of our right brain is designed for patterns. When we look at an incomplete or muddled picture, for example, a Polaroid blurred by time, our brains fill in the missing elements to recognize with ease the identity of a loved one. Another key component of the right brain's unique skill set and attribute is the ability to feel the experience of someone other than oneself, meaning through empathy.

Our perspective of "other"—our ability to discern the motives, feelings, and intent of other beings—is of critical importance and primarily driven by the right hemisphere of the brain. Even our "cradling bias," the strong universal tendency to hold infants so their gaze directs to the left side of our face (the right hemisphere's visual field), suggests that emotional qualities such as love, nurturing, and empathy are primarily mediated by the force of the right-hemispheric brain. The right hemisphere is so important to the maintenance of our relationships that its impairment may even result in an absolute rejection of oneself, as we often see in patients with the condition of hemineglect.

So how does all of this affect our understanding of God?

Theory of Mind and the Reflection of Faith

> The aim and purpose of human life is
> the unitive knowledge of God.
> —ALDOUS HUXLEY[11]

What Huxley calls "unitive knowledge" points us to God because our minds were designed in such a way as to acknowledge and awaken to our purpose in the universe, to discover this hidden order.

Evidence from a variety of clinical and neuroimaging studies of the frontal lobes during feeling states has revealed increased activity associated with particular feeling states, such as empathy and awareness of others. These specific areas of the frontal lobe brain systems are also active when people weigh their actions within a moral context. The point is that if we were merely biological, then we would not have developed empathy for others. We would have no evolutionary reason for "filling in the blanks" and extrapolating the existence of God.

Empathy is arguably the single most important as-

pect of larger faith. Since we have empathy, we understand that the universe is not simply about us—and that
understanding points to something "beyond" ourselves.
If we were merely biological, then we would be interested
in only our own interests. But because we are both biological and "beyond biological," we are also interested in
the interests of others.

Our capacity for empathy requires belief, a belief that
all of life, from the smallest to the greatest, has purpose
and meaning—an existence that may be beyond our limited perception yet still within the Mind of God. This
means that although belief and faith are experiential and
subjective, their concern is always about relationship and
reciprocity. We experience God's love when we exhibit empathy, and we can glean God's compassion when we experience compassion. We reveal God's mercy when we act
mercifully, his kindness when we are kind, his love when
we love. And our brains make this possible. We can show
this "reflection of faith" even more clearly through the
study of the action of mirror neurons, a major finding in
neuroscience research.

Mirror Neurons and the Reflection of Faith

Italian neuroscientist Giacomo Rizzolatti and his colleagues at the University of Parma first identified mirror
neurons, the discovery of which helps explain how and
why we "read" other people's minds and feel empathy for
them.[12] Essentially, these specialized cells in our brain are
activated the same way when we see something as when

we do it. Not only do our brains "mirror" what they see, but they also provide us with a reflection (no pun intended) of the nature of our being.

This activity is vastly important. It shows how our social relationships have a shared understanding and intentionality, a collective space, and how the ability to find common ground is central to forming relationships. Our mirror neurons are used to discover something in the mind of the person we are watching or relating to, and this is an essential part of socialization. Mirror neurons allow the embodiment of mind, the immaterial experience from within. It is through the activation of mirror neurons that our brains are able to make the mental pilgrimage from the objective to the subjective, from the conditional to the unconditional. Mirror neurons are nature's way of giving every person the ability to take "metaphysical selfies."

For instance, if you and I are in the kitchen together and you accidentally touch an element on the stove and burn your finger, I will undoubtedly wince, too. What has happened just then in both of our brains is that when you touched that hot element, certain neurons fired in your brain. Then, thanks to mirror neurons, certain neurons in my brain fired almost simultaneously and produced similar results of pain. I didn't feel the exact same pain you did, but I felt a similar pain, a vicarious or shared-experience pain, and I could feel this pain both because I've felt it before from my own experiences and because I see pain in you and I care about your feelings, too. It's like my neurons are looking at your neurons and "mirroring" them. Mirror neurons teach me something about myself. They also teach me something about you.

Of course, there is a gap between facts and meaning, between action and intent. How do our brains, through the activity of mirror neurons, transcend the raw data of our sense of the universe and its inhabitants and process these experiences into something that has value and purpose? This process, whereby mirror neurons not only reflect but help interpret the meaning of behavior, is referred to as theory of mind.

Theory of mind describes our brain's unique ability to appreciate and acknowledge the reality of our mental states—that our beliefs and knowledge are real—and to understand that others have beliefs and knowledge that are equally real.[13] Theory of mind includes the capacity to put ourselves into someone else's shoes to imagine what it would feel like to experience something from a perspective totally not our own. This perspective relies on the belief that others have an invisible mind.

Being able to attribute mental states to others also requires an understanding that the mind (by its very nature) functions as a generator of representations. It's through these representations that we know the meaning of love or the experience of pain. It's through mirror neurons and their capacity to generate a theory of mind that we can imagine and believe what is otherwise unverifiable. In essence, our ability to reflect on another's experience as if it were our own is the capacity for empathy.

When groups of specialized neurons in our brains are activated by our experience of another's emotional state, we are in a sense embodying the other person. By occupying the anatomical space of mirror neurons, our awareness of others is no longer theoretical—it is a state of organic "betweenness." On the basis of mirror neuron activity, empathy is the capacity to embody another person. It is through the brain's capacity for empathy that relationships are made possible and our deepest connections thrive.

Why is this so important? Mirror neurons help us to discover something going on in another person's mind. So if I can discover the feeling of pain in your mind when you touch a hot stove, then (as long as a relationship exists) I can theoretically discover something in anybody's mind.

And that could be why qualities such as compassion exist. Remember, there's nothing within a strict evolutionary framework that should produce compassion within people. If life is only about the survival of the fittest, then there is no compelling biological reason why we should become compassionate people, or even know what compassion is. Certainly we might have bound ourselves together as early humans to protect ourselves against some greater danger, but that still doesn't account for the root character of empathy. A band of people is not necessarily an empathetic group of people. Yet we still have evolved to have this quality of compassion today. Why?

It's because "God" is compassionate, and our brain's mirror neurons went to work and mirrored this same feeling. We feel compassion today because God first is compassionate. All of the intangible qualities such as hope and love and joy and compassion are found within the Mind of God. We become aware that our actions are a reflection of a deeper reality, as it is written in Proverbs, "As water mirrors a face, a heart responds to another."

Evidence from Above

Regarding my personal spiritual development, one of the aha moments I had regarding the existence of a Higher Being happened a few years back when I took my two young daughters, Julia and Sofia, to the Hayden Planetarium at the American Museum of Natural History in New York City. We settled into our seats to await one of the "immersive environment" shows in the auditorium, and Sofia said to me, "Dad, I know why God created us." I asked why. Sofia added, "Because he was lonely."

From the perspective of neuroscience, I knew what Sofia was getting at: the idea that God could want something else in the universe with a brain like his, a brain that could mirror what he felt.

I was still thinking about Sofia's comment when the show came on. Astronomers have been observing a web of filaments, an immense network extending millions of light-years through space—including through our Milky Way—that imparts the large-scale scaffolding for the structure of the universe, but also perhaps its communication network, which we can't yet fathom. These observations suggest that galaxies in the universe are neither disconnected nor isolated by vast expanses, but form interconnected parts of a complex macro-architecture designed for a particular function—connectivity.

The image projected on the hemispheric dome overhead struck me as so familiar that it felt as if I were experiencing déjà vu. As a neuroscientist, I have looked at a brain under a microscope countless times. It dawned on me just then that it looks almost identical to the universe! The appearance of the web of the universe bears an uncanny—almost a spooky—resemblance to a cross section of the brain's anatomy, packed as it is with networks of stars resembling neurons and galaxies as extending dendrites and synaptic junctions designed for connectivity, communication of information, interrelationships, and, of course, consciousness.[14] Looking at the glimmering web of cosmological star dust against the unfathomable darkness, I could not help but imagine this image as the Universal Mind.

My aha moment was this: Our brains are constructed in such a way that they give us an image of the Mind of God. We can divine from the remarkably parallel structures of the universe and the human brain a message about relationships of being-as-bonding, not of partition, not alienation, segregation, or isolation. Theory of mind and the inherent capacity of our brains to experience empathy are the cornerstones for our belief in God. Empathy is essential to establish connectivity within us and between us. It's at the nexus of this engagement where we are apt to also discover the Mind of God. When my daughter Sofia contemplated the Mind of God as we observed at the planetarium an otherwise unimaginably vast expanse of cosmological emptiness, she inquired about God's perspective as only a child could do. She had "empathy" for God by building a representation in her own mind of what possibly is in his mind.

Knowing the Mind of God requires a shift in our

own way of thinking—from being a "theory" or belief to becoming a primary and fundamental axiom of mind, where to acknowledge God means to acknowledge the other within ourselves. Each existence comprises a universe contained within a brain; in each, its own history, activity, and destiny. And within all of us, this Mind of God acts as a conduit to every other and to the immeasurable "unknown." Each of these extant brain-worlds is composed of a microcosm of the larger universe: the universe within. Our brains and the universe beyond are tethered together as ships are to moorings, as mind is to body, and as soul is to Source.

We Are So Close

We can acknowledge God by understanding that there's an existence outside our material being. The right hemisphere of our brain provides strong evidence of the existence of the other. Is this definitive proof that God exists? No, although it acknowledges an ongoing inquiry whereby the answer is ready to be discovered once we begin to ask the question in earnest. Perhaps a concluding anecdote can explain this best.

There once was a private investigator hired to confirm a husband's suspicion that his wife was carrying on an affair. The private investigator followed the man's wife and discovered she was having secret lunch meetings with another man. He continued to follow her and soon confirmed that the couple was renting hotel rooms. The private investigator reserved a room across from theirs. He set up a telephoto lens and focused it on the illicit tryst. But just before he could snap a photograph, the woman and the man closed their blinds. Frustrated—yet very close to being certain—the private investigator concluded the case by commenting, "We were so close to getting definitive proof."

This is our lot as well. God exists when we "act as if" he exists. We live in an inescapable subjective reality. Neuroscience reminds us that when we see an image of the external world, what we are really seeing is only our brain's own particular reconstruction of light that has traversed through our retina and been projected onto the visual cortex. There is an almost infinite gap between the structure of our brains and the universe beyond us, between the "what is" and what we believe to be reality.

We believe based on what we experience of the world: what we see around us, what seems to be, and how we choose to respond to the resultant meaning we uncover from such an experience. Our capacity to "believe" requires us to employ extrapolation, a version of inductive logic in which we will, cardinally, never have access to a complete and fully informed picture. When we believe in something, it means that we've opened our mind to the capacity to fill in the "illusory contours" of reality, intuiting from the visible signs of life the possible—or even likely—existence of an invisible order. We open our mind as this hidden truth discloses itself to us. This is perception of presence in ostensible absence, and it's not so hard to do. Our brains are particularly apt at filling in the blanks of experience formed from partial sensory input. Without that facility, we would never have discovered anything, invented anything, or evolved into anything.

Our understanding of God is inextricably linked and fully dependent on our biological nature. Our brains are designed for faith. Faith is an integral component of the human experience, and our brains evolved with this capacity for a higher purpose. Faith calls us to inquire not only about facts but also about their meaning. This purpose is for relationship—relationship between each other and between ourselves and, perhaps, an unseen Creator. Relationships require empathy, understanding, and genuine acknowledgment of other. Our left brains are inclined to see relationships based on their utility to ourselves, whereas the right brain sees our individual existence as part of a much larger interdependent landscape.

Without empathy, a fundamental tenet of almost every religion, there would be no social cohesion, cooperation, or regard for anything other than self. Our right brain and its associated mirror neurons have evolved not only to objectively know things but also to have wisdom about them. Our belief in God is fundamentally the awareness of a universe filled with other sentient beings and perhaps a Higher Being that transcends this, too. We can "know" the reality of this existence by sensing it, by feeling it—through the experience of it, and in this way our right brain provides a path to this deeper level of understanding than we readily can see or measure through our limited senses.

It is through our lives—our beliefs and actions—that we manifest God's otherwise unknowable reality.

The Neuroscience of the Soul

And now we might add something concerning a
certain most subtle Spirit, which pervades and
lies hid in all gross bodies; by the force and action
of which Spirit, the particles of bodies mutually
attract one another at near distances, and cohere,
if contiguous; and electric bodies operate to
greater distances, as well repelling as attracting
the neighboring corpuscles; and light is emitted,
reflected, refracted, inflected, and heats bodies;
and all sensation is excited, and the members of
animal bodies move at the command of the will,
namely, by the vibrations of this Spirit, mutually
propagated along the solid filaments of the nerves,
from the outward organs of sense to the brain, and
from the brain into the muscles.

—SIR ISAAC NEWTON[1]

Do we actually have a soul?

I'm not talking about the "soul" of music, or the
"soul" of a strong community. I'm talking about the literal immaterial part of a person that contains spark, life,
feelings, thoughts, and "mind."

The question provokes much passionate debate. The halls of philosophy are littered with the corpses of arguments for and against the soul's existence. Philosophers have turned to science to break the stalemate, and as neuroscience lays bare the inner workings of the brain, it has much to contribute. Science has even more to tell us about the soul than we've yet suspected.

Philosophers of religion generally affirm the existence of the soul. But in mainstream philosophy today, as in science, the reigning view is materialism. Materialists believe that there is no soul per se; we have merely our bodies or matter. For materialists, all other positions are put in the category of fantastic literature or religion and are thus suspect. As materialist philosopher and cognitive scientist Daniel Dennett says, human beings have a "soul. But what's [the soul] made of? It's made of neurons. It's made of lots of tiny robots. And we can actually explain the structure and operation of that kind of soul, whereas an eternal, immortal, immaterial soul is just a metaphysical rug under which you sweep your embarrassment for not having any explanation."[2] Dennett's soul really isn't a soul at all. It's a switchboard. Many scientists agree.

As a neurologist, I agree that the workings of basic biology and matter have much to explain concerning the soul, but I disagree with the view that reduces all mental phenomena or consciousness to material actions. Any view that reduces our mental states to mere neuronal, electrochemical reactions, that sees the brain as a mere computer, is sadly inadequate. As we'll soon see, advances in neuroscience show that our spiritual states transcend material brain structure and function. New evidence suggests that there may be such a thing as *mind* and soul.

Neuroscience can also show that our belief in the existence of our soul is essential to our being. The question of whether we have a soul is not an idle one; it is one of the most important inquiries we can make. One might even say that evolution's highest goal is the belief in the soul, for that belief not only provides the foundation for our inquiry into the meaning and purpose of our existence; it informs our actions and provides our life with its highest moral dimensions.

What Exactly Is the Soul?

How can neuroscience prove the soul's existence? The evidence is found in what we already know about the way that our biology—our brains and our minds—works. It's found in what we know about a striving for connectivity, meaning, and relationship that is deep within our makeup.

The soul at the simplest level is an extension of the body's striving for connectivity. Without its ability to bond, the soul is otherwise remote and inherently unknowable, even unto itself. The soul reveals itself through its will and engagement, an engagement that recognizes that its vitality and expression are contingent upon its relationship with that which is beyond itself.

This is not a new idea. Some of the earliest views of the soul affirmed its relationship with the body. Many early thinkers saw the soul as an animating spirit, one that enlivened the body but was not totally "of" the body.

To the Greeks, the soul was *psuchê* or *psykhe,* literally the "breath of life." For Thomas Aquinas and Aristotle, the soul, or *anima,* was that which animated the body but was not exhausted by it. Isaac Newton saw the soul as energy, breath, and Spirit.

Newton added scientific insight into how our bodies actually work that grounded and gave weight to the view of the soul as the life force. His description of the soul as energy still rings true. The soul is the entity that maintains the vitality of our body; it imbues life to our organs and limbs. The soul is also that which gives breath to what he called "intellect" and to what we'll call "mind"— the power and will to create, understand, and relate. It is our mind/intellect that gives us both a sense of selfhood and through that individuality a way for us to connect and maintain vitality beyond the body.

One note on terminology. I use the words *mind* and *soul* fairly interchangeably. As discussed, the crucial distinction I want to make is between the material (body) and the immaterial (soul). The soul cannot be reduced to body, and something more than mere body exists. But whether we call that something "mind" or "intellect" or "soul" is less important to the overall argument than understanding how the mind/intellect/soul transcends the purely physical processes while remaining connected to them.

How does Newton's Spirit imbue our bodies and souls with life? Newton believed the Spirit possessed action or force. Through its force, the Spirit excited sensations, compelled the mutual attraction of bodies, and even emitted or refracted light. I argue that that force is our striving for connectivity. And this striving is fundamental to understanding the biological basis of the soul. But before we delve more deeply into the biological basis of our fundamental connectivity, first I'd like to tell a personal story that illustrates something about what that "striving for connectivity" looks like. It also shows why talk of the soul—which, I'll concede, does involve neurons—should not be limited to measurable physical elements. If we discuss soul, then we must examine not just strivings on the level of biology but also our deeper strivings for connection. In fact, the parallel between our bodily strivings and our spiritual ones is at the heart of my reason for believing we must have a soul in the first place. More important, if we reduce all talk of soul to body, we have irretrievably lost an important element of who we are.

When I was nineteen, my father went to visit my brother in California. While there, my father had to undergo heart surgery and suffered a massive stroke in the operating room. This complication is more common than most people think. As a result of the stroke, Dad slipped into a coma and was unresponsive for days. My family feared the worst. When I heard the news, I immediately arranged to fly to California to be by his bedside with my brother and the rest of my family. Before I left the East Coast, however, there was something important—and unusual—I felt compelled to do.

I grew up in a mostly nonreligious family. Though we considered ourselves Jewish, the only exposure to religious Judaism I had was going to friends' bar mitzvahs—

not exactly a spiritual experience for most Reform Jews in America. But while in college, I met Rabbi Gurary, a beautiful and gentle man of unwavering faith, and he became a friend, mentor, and spiritual teacher. Rabbi Gurary came from the Lubavitch tradition, a sect of Hasidic Judaism with its roots in a region of Russia not far from my paternal origins. *Lubavitch* means "Town of Love," and I learned from the rabbi that to love our fellow humans and to love God is one and the same process, one and the same obligation.

Before flying out to see my father, I implored Rabbi Gurary to ask the Lubavitcher Rebbe—the Grand Rabbi of the sect, widely revered by millions across the planet—to say a blessing for my father. This would be a massive honor, and a long shot: Rabbi Menachem Mendel Schneerson, the seventh leader in the Chabad-Lubavitch dynasty, was considered the most extraordinarily influential Jewish personality of modern times (he still is, though he died in 1994). Again, I wasn't particularly religious, but I'd heard many stories of "miracles" following the Rebbe's blessings—and I wanted one for my dad. While my rational, logical brain knew how serious my father's physiological situation was, I left a little door open to such a miracle, and I knew a little prayer—or in this case, a big one—couldn't hurt.

When I arrived in California, I found my sister, mother, and brother shuffling between the ICU and my brother's small house, where we all camped and kept vigil. As is common in crisis situations, family members wept, stared into space, snapped at the slightest infraction, or

kept themselves overly busy with necessary and unneces-
sary errands. My mother had neither eaten nor slept for
days. She looked to me like a wingless and featherless
bird whose bones might snap if she fell out of the nest. I
had never seen that strong woman so fragile before.

My sister had warned me about my father's appearance. I told her I could handle it. But she cautioned again
that nothing prepares you for seeing someone you love
entangled in IV lines, bloated, hooked up to a respira-
tor, and unresponsive to anything you say, even to your
touch. She looked at me a long time, and I could tell she
was trying to convey the message that she didn't think
our father would ever come out of his coma. For all in-
tents and purposes, I was going to visit a dead man.

I held my breath, hoping that the Rebbe on Eastern
Parkway in Brooklyn had received my request for a blessing in my father's name. I didn't tell anyone in my more
or less irreligious family of this secret wish or about the
inklings of my emerging faith.

When we got to the hospital, my mother finally
reached her breaking point. Afraid she would collapse
from exhaustion, we helped her into a chair in the waiting lounge, and my sister stayed with her. My brother
went to get her some much-needed sustenance from the
cafeteria. So I unexpectedly entered my father's room by
myself.

My sister was right. My father looked as close to dead
as I imagined a human could look outside the coroner's
workplace. His eyes were stuck shut; his beard was half-
grown; and his body was twisted into an abnormal pos-

ture. I didn't know what to do. I whispered a feeble hello to him and rubbed his hot hand. Then I said one word: "Dad." I said it again and again. Then I added, "It's your son. It's Jay! If you can hear me, open your eyes. Or give me a signal. Show me you can hear me. How about you squeeze my hand? . . ."

I worked on the hand squeeze for a while, kneeling by the bed and staring intently at his fingers for any sign of the minutest movement. For the longest time all was silent. Then, out of the corner of my eye, I became vaguely aware that something in the bed above had changed. Slowly I looked up and saw that my father had turned his head toward me and opened his eyes—for the first time in days. He seemed shocked, and he looked frightened. I, too, was scared. We were like two animals coming upon each other in the forest at night, neither knowing what to do.

Underlying the shock and awe that my father's gaze was now directed at me with full force was a sense of profound connection and emotion. It was the kind of deep connection many fathers and sons have; but more than that, it was a connection that in its very strength frightened us. It was as if in speaking to my father, I had called to God, and God had answered by having my father open his eyes. Was that even possible? I quickly rushed out of the room and called for the hospital staff. Then I bolted to the waiting room to tell my family that Dad had come out of his coma. This is not a message one gets to deliver every day.

Utter disbelief seemed to reign in all my family

members and even in the staff as a nurse escorted my mother into the room. My mother half smiled and said my father's name. I didn't follow her to the bedside. Instead I stayed by the door, then found myself walking down the corridor in the other direction, trying to make sense of what I had just witnessed. When I saw a red exit sign above a set of fire stairs, I made a beeline for it—but I never made it that far. My legs stopped carrying me, I crumpled to the floor, and I cried in a way I hadn't since I was a child. I felt undone. I'd experienced a power greater than mine, yet it was all too much to take in.

In the years since then, I have often thought of the connection my father and I experienced that day. The indescribable awe and gratitude to be touched like that by an unseen hand led me to feel an immensely strong and mysterious connection, a unity, deep in my marrow. Many people had uttered the same words to him: "Give me a sign." But only when I said it did his eyes flutter open. I had the palpable sense that my father and I were both striving to reach each other but also that we were connected *beyond* each other. In that brief moment, the connection was both to my father and to God. It was a unity of our souls, and in our souls' unity there was a connection to and communication with something higher, a greater communion with God.

Yet what concrete evidence do we have that souls exist and strive? That our souls are driven to connect with each other or to commune with God? That they are driven to the kind of unity my father and I felt with each other—and even with something higher? If the soul is not corporeal, if it is not biological but perhaps "meta biological," then how can what we know about the body, about the brain, help us understand whether we have a soul at all?

The Brain's Drive to Connect

Consider our basic striving for connectivity. The drive to connect is deep within our biological makeup. The brain is an organ that is constantly striving for relationship for that which is beyond itself. We have all kinds of drives: survival-induced biological drives such as the drives to eat, drink, have sex, and stay alive. Whenever there is an imbalance in our biology, our physiology takes over and we seek what we lack. If we do not have enough water, for instance, we thirst and are psychologically and physically motivated to find water and drink. The body constantly works to maintain a state of equilibrium, or "homeostasis," in many of its internal physiological processes.

As per the laws of thermodynamics, all living biological systems seek their most stable state. The brain similarly has a tendency toward maintaining equilibrium.[3] To achieve this equilibrium, the brain is driven to seek certainty and finds its most stable energy state when it can accurately predict what is happening based on patterns that it already recognizes. The brain's drive toward certainty—to find order in a complex and often chaotic universe—is what leads it to create predictions and beliefs; that is, through predictions and beliefs it can generate a predictable result.[4] But how does it generate those predictions and ensure certainty? This is where "mind" enters.

Mind and the World of Ideals

The brain is comparable to Plato's "sensible" world of matter, the physical plane of existence, whereas mind constitutes the "intelligible" world—the world of ideals. Mind is our capacity to entail representations of scenarios; it is a sort of "life simulator." If the brain's drive to connect is a relatively simple and brute biological force, then the mind's drive to connect is, perhaps appropriately, a much more sophisticated and powerful tool based upon more abstract notions of reality, language, imagery, metaphor, and religion. For example, through language we generate representations, models of reality we construct and generally agree upon. We agree that at this thing called a store, I will exchange these green pieces of paper we call money, the value of which we've already agreed on, for goods that you want to sell and I want to buy, and that the transaction will take place without force but potentially with much persuasion.

Language enables us to construct ever more complex beliefs and to rely on those beliefs in our daily lives. All of which is very necessary given the complexity of our world. Language is what makes possible our sophisticated monetary-based system of exchange where everyone understands what something costs and what they are exchanging for it, whether they are paying with green paper, a piece of plastic, or even a phone. The construction of complex beliefs that we can rely on through language also helps us deal with and understand the complexity of other human beings. Language helps us sort out conflict. We talk things over and reach agreement. Through language, we trust that our agreements will be honored.

Religion is another sophisticated tool our minds have developed to connect and establish trust. The same mechanisms of belief regarding more mundane matters apply equally to more existential ones. Religious beliefs seek to generate predictions that reduce uncertainty. They create models of reality and models of behavior we can agree on. Ideally, religious beliefs generated by our minds reflect empathy toward our fellow human beings and allow for relationships based on intimacy, connection, mutual respect, and purpose.

How do we actually know if what we believe is real? Based upon the representational nature of reality, what we believe becomes our truth. The mind can only know its own version of events.

The kicker is that the constructs or patterns that our brain has generated don't necessarily have to be truthful ones. We still believe them. But belief, as it turns out, is not just immaterial or ephemeral. Mind has a direct influence on our physicality. A direct bridge spans our physical states and our psychological ones, the function of synapses and the experiences and perceptions they enable. Beliefs exert themselves on our material being. What we believe—and what we don't believe—has a powerful effect on our bodies. A case at the very beginning of my medical career illustrates how our health, even our very existence, is utterly contingent upon what we believe.

When I was a young intern, a near-term pregnant Caribbean woman, Celestine, came into the ER complaining of respiratory distress. Her medical workup was completely normal, and the staff concluded she was having a run-of-the-mill anxiety attack, possibly related to

her upcoming delivery, a reaction that is not at all uncommon. So they ordered a psych consult—the standard
of care—and I met with her. I soon learned that Celestine was emotionally devastated by the unfortunate fact
that her pregnancy (she was eight and a half months
along) was the result of an illicit affair she had had with
a married man, a native of West Africa. He had tried to
force her to abort the baby earlier on in the pregnancy,
but she had refused on religious grounds, insisting on
carrying her child to term. He ended their relationship
and repeatedly threatened that if she didn't abide by his
will, he would cast a "hoodoo spell" on her to kill her and
curse her baby, she said. I could tell as she recounted this
threat that her blood pressure and respiration rate rose.
Celestine believed she would die, she told me, but she was
determined to have her baby anyway.

"Don't worry," I assured her. "You are just experiencing anxiety. All of the tests we conducted came back normal."

"Don't you understand?" she replied. "It's a curse and
I am going to die."

It was nearly impossible to keep Celestine calm. Her
heart rate soared all night. She went into early labor the
next morning, hyperventilating during her contractions,
unable to catch her breath despite the administration
of oxygen, and repeating the whole time, "He killed me.
Please save my baby—please." Having done a full workup
on her only twelve hours earlier, we concluded there was
no medical reason that she should be suffering these
symptoms. The medical staff again attributed her com-
plaints and symptoms to anxiety and kept trying to set-

tle her nerves. Celestine delivered a healthy little boy, but soon thereafter she developed respiratory arrest. Despite herculean resuscitation efforts, she died without ever getting the chance to hold her child.

Beliefs are so powerful that they can shatter our reality and replace it with an alternative one. Indeed, the virtual realities that we create for ourselves can be all-encompassing. Celestine shows us that it's not all in our head; our beliefs in turn powerfully affect our body.

The Ability to See the Mind

If this mind/soul actually exists, then where in our biology does it reside? Fortunately, the "gaps" between the physical nature of our brains and the nonquantifiable attributes of mind are getting smaller and smaller with each passing year and every new discovery. Think of your brain structure as the computer hardware. The neurons, the glia, are the anatomical structure where activity takes place. Traditionally it's this hardware alone that we've been able to see and diagnose. But thanks to new imaging tools, we can also now quantify to some extent not just the hardware but the software. Nowadays we can use diffusion tensor imaging (DTI), a magnetic resonance imaging (MRI) technique by which doctors can peer inside the brain and see what is happening—the consciousness or spirit, if you will, within a person.

DTI, by measuring the flow of water through the myriad networks within the nervous system, allows us to measure free energy in the brain, in order to create a visual image of the "soul in action." We can literally see images of the brain's synaptic map—its highways, its bridges, and its detours. We can spot neurological disorders in ways not previously available to us.[5]

But most important, we can now visualize the Spirit to which Newton referred: the dynamic forces that move the energies of the brain and that in turn direct our limbs and organs and the rest of our body. We can quantify that energy. Now with MRIs we're not just looking at neurons and glia; we're looking at activity. And this gives us more of a glimpse into what the mind does. We can literally see the "mind" in the brain. We can see belief in action. And we can see how various mental states (spiritual activity or mind software) change the structure of the hardware.

The Bridge Between Mind and Body

Our emotional states or beliefs affect our biology. But *how* do our thoughts and beliefs inhabit our physiology? In other words, assuming our brain is a virtual reality generator that generates perceptions, *how* do those perceptions *interact* with our physical processes? How do nonphysical events or metaphysical states interact with the physical structure of the brain? How does mind/soul impact our body? In order to understand how our perceptions or beliefs interact with the physical structure of our brain, we must understand how our mind/soul functions as a hologram.

In the 1930s, pioneering Canadian neurosurgeon Wilder Penfield first made the connection between the hologram and how the brain projects its virtual reality.

A hologram is a three-dimensional picture produced by a singular source of light. The word *hologram* derives from the Greek words *holos,* which means "whole," and *gramma,* which means "message." To create a hologram, we take a picture of an object, say, a tennis ball; then we illuminate the plate and look at it from various angles. The result is a realistic 3-D image we can observe from all sides. The diffusion of light produces the virtual model. But, of course, the tennis ball shown in the hologram itself doesn't exist. It doesn't occupy physical space.

Penfield developed a surgical procedure in which he operated on patients' exposed brains while the patients remained fully conscious. The patients reported precise locations in their bodies where they felt sensations elicited by Penfield's electrical stimulations to their brains. Ghoulish though it was, it wasn't terribly surprising they could feel his probings. Yet as Penfield used the tip of his electrode to chart the function of the cerebral cortex related to corresponding locations on the body, he elicited in his patients something no one expected: three-dimensional dreams, smells, auditory and visual hallucinations, long-lost memories being vividly recalled, and even powerful out-of-body experiences. For instance, according to Penfield, "When an electrode was placed on the patient's temporal lobe, the patient had a complete flashback to an episode from earlier in his life. They were electrical activations of the sequential record of consciousness, a record that had been laid down during the patient's earlier experience."[6]

Penfield concluded from these experiments that areas

of the brain were able to enfold memory in a holographic way, providing a dimension of perceptual reality that allows us to *re*-animate, and thus re-experience remnants of our prior experience. When you remember your mom soothing you when you were a child, what you experience in your mind isn't a flat visual like a photograph or a painting. Rather, your mind tends to re-create the actual event. You see your mother's face in a three-dimensional way. You rerun in your head a movie of her movements. Your memory of your mom is closer to a hologram than a photograph. Which is exactly what Penfield found. When the brain was externally stimulated, it produced a compelling reproduction of prior experience. That reproduction represented the remembered event in its full multidimensionality.

Penfield's patients' flashbacks seemed to play themselves out in proper order like scenes in a movie, as though there was a continuous flow of events. He noted that patients were able to distinguish between their current experience in the operating theater and the vivid memories they were recalling. He might have been able to ask them, "How are you enjoying the movie?" But they wouldn't have quite understood. In other words, they were experiencing two simultaneous "streams of consciousness." On the one hand, they could see and feel and experience the memory, fully revivified. On the other hand, they also knew it was merely an observation. Penfield's subjects knew they were generating a virtual world at the same time that they remained anchored in actual reality. They could even compare the two. Penfield's dual-stream patients led him to the conclusion that the mind was indeed separate from the brain and was supported by a "different form of energy."[7] Numerous subsequent studies back up Penfield's findings.

What the hologram analogy tells us is that the mind doesn't occupy a clearly delineated physical space within the brain's hardware. It is non-localized, with all the separate streams of thought inextricably linked to other clusters of recalled experiences and to the whole. This flies in the face of conventional thinking in neurology, where we generally associate function with specific delineated areas of the brain. But the hologram example tells us that whereas the brain is physical, the experience that it embodies—its virtual or representational reality—is immaterial. It's as if the hologram enables a transmutation of sorts in which the invisible forces of the mind and the soul are made manifest through the vast anatomical network of neurons and synapses. The hologram connects mind and brain by providing the nexus between the material and immaterial, between the physical and metaphysical.[8]

The Bridge Between Mind and Soul

Viewing the mind as a hologram helps us understand how perceptions and, hence, beliefs are generated by the brain and how immaterial thoughts and emotions interact with our physical processes. The challenge with the holographic model comes when we try to understand what the brain is actually perceiving. Going back to our projected tennis ball, the ball in a hologram isn't there any more than are the shadows on the cave wall in Plato's *Republic*. The hologram is a trick of perception—it's supercondensed, realistic information, recorded and re-animated. When we look at the hologram, we see our tennis ball projected in space, and the ball shifts in perspective as we view it from different angles. The holographic model implies that our perceptions are merely an illusion. If we are perceiving a simulacrum, what is the true nature of what we are perceiving? What is the nature of the thing being reflected?

There's an important paradox here. It is one thing to assert that the hologram is the conduit between thought or emotion and body, between our memories and our physicality. But if we know that our brains are capable of projecting illusions, then we will also want to know what is behind the screen of that virtual projection. We will want to know that the hologram is real—or at least that what is represented by the hologram is real. We want to know our beliefs are real. We may be pulling the blinds off the magic trick only to conclude that there is nothing behind the curtain. Or is there?

Remember that the hologram is an illuminated image that requires a light source to create it. Without light, there is no object, nor a simulacrum of it. This light, too, kindles the spark of human consciousness—our ability to perceive. And just as light is always connected to its source, our souls are the "objects" connected to the emanated light of God, illuminating its very existence. "Light disperses its unitary nature into an infinity of forms," the eighteenth-century German philosopher Hegel tells us in 1807, "and offers up itself as a sacrifice ... so that from its substance the individual may take an enduring existence for itself."[9]

The soul, then, is our connection with God, our conduit to a deeper reality that is otherwise unknown to us. Believing in the soul means believing that there is an immaterial drive analogous to the material drive (that is, to the brain and mind's drive) for connection that is deeply embedded within our makeup. That there are parallel forces of energy, if you will. This drive for connection is what leads us to seek our intrinsic unity and attachment.

We are driven to connect to God, just as a flame rises upward to embrace its source. Descartes saw the soul as establishing not only what is *not* biological but also what *is* the spiritual principle of human beings and the "spark of Godliness" within. We are all born with souls, with indivisible essences bound to God, but there are multiple levels of divinity revealed to and concealed from us. The highest level of communication or communion connotes the essence of the soul—the soul's unity with its source, the singular essence of God. The soul can be regarded as the focal point or link between our engagement with the physical world at one extreme and with God at the other. Each soul can be regarded as an integral fragment of divine light,[10] the essential spark between a transcendent God and his immanent manifestation. The soul is what connects us to God, to each other, and to ourselves.

Although my father eventually came out of his coma, he never fully recovered from the stroke. The resulting emotional pain tested both his and our family's faith and fortitude over and over again for the next twelve months.

Almost exactly a year later, he died from complications from the stroke.

My father had been a highly social, verbal, and witty man—a podiatrist who joked with his patients, cared for them deeply, and got to know them intimately. In the last year of his life, the stroke left him with a significant speech impairment, blindness, and partial paralysis on the left side of his body, not to mention a kind of hollowness where his personality used to be. I was continually struck by the idea that he came out of that ICU a still living man, yes—but a different, clearly deficient man.

My father's profound disconnection during that period (he was unable to speak or clearly recognize on a physical basis who his loved ones were) affected me as much as our moment of connection had earlier. I was terrified and shaken by my inability to access the man I had always known. What had happened in his brain and his mind to change his disposition so dramatically? Whatever had been shut off in my father awakened in me a great thirst to plumb the mysteries of the mind. I decided to study neurology and neuroscience. I became fascinated by our minds'—and our brains'—inner workings. The disconnection with my father provoked an entirely new path and career for me.

Together, our moments of disconnection and connection propelled me forward. Yet for me, it is primarily the moment of electrifying connection that has triumphed and stayed with me. It was a moment during which something within us connected not just to each other but also to a higher power. A moment driven by something deep within the makeup of each of us. A moment that drove me forward in my scientific and religious pursuits.

I suggest that something similar can happen to anyone, when our drive to connect to each other is interrupted by a profoundly traumatic experience. Not only can war, criminal acts, anything that pits a soul against another soul result in a huge loss of human life; it can also lead our souls to cease seeking connection and to focus on mere survival. Our deepest trauma is first and foremost the loss of trust in ourselves, and in parallel, the loss of trust in others, and our loss of trust in God. In the Hebrew language the word *sin* also means "to diminish"; our separation from God and from one another is a departure from our essence, which is fundamentally one of attachment and connectivity.[11]

Yet there is hope. As we've seen with our discussion of the brain and mind, we can re-establish this connectivity when we elevate our existence beyond its physical contours. When we reach beyond the physical to the metaphysical, we rediscover our connections to one another. The purpose of the brain/mind's creation of belief is to endow us with trust, the predictability of experience, and the cohesion of relationships. It is the foundation of our connectivity within and among ourselves, the capacity to act in a world where reality is otherwise regarded as purely subjective. It is through our ability to hold the intangible that we are able to also grasp the essence of ourselves and each other and, through this connectivity, cleave to God as well.

The Evolution of Faith and Reason

The limits of language mean the limits of my world.

—LUDWIG WITTGENSTEIN[1]

By creating symbols, the mind comprehends what is in itself incomprehensible; thus in symbol and adage, the illimitable God reveals Himself to the human mind.

—MARTIN BUBER[2]

Let me introduce you to a close friend of mine.

I love this friend. I truly do. My friend is an embodiment of acceptance and love and fun, and over the years he has become a trusted member of my family. Let me see if I can describe him better.

He has brindle coloring and a head as big as a basketball. He loves my wife and children and is highly protective of them. When people first look at my friend, sometimes they are a bit afraid of him. He's all muscle, even though in his heart he's all mush. He wouldn't hurt a fly. He weighs about ninety pounds, is altogether beautiful, and answers to the name Tiger.

Who's my friend?

He's my dog, part pit bull, part mastiff, a breed combination that can sound shocking until you see him and
get to know him. Originally we rescued Tiger, who was
abandoned when he was about a year old, found emaciated and homeless. I wanted to have a watchdog in the
house when I wasn't around. But Tiger has proved to be
the biggest scaredy-cat in the history of dogdom. He's not
a guard dog so much as he is a guard kitten. His idea of a
wild and adventurous time is to ride in the front seat of
the car when we drive.

I mention Tiger because as much as I love my dog and
as much as he's a valued member of my family, I want to
surface a bigger question: Is there a difference between
Tiger and the rest of my family? Is there a difference between Tiger and my eleven-year-old daughter, Sofia, an
aspiring magician and keen observer of human behavior, a devoted hockey fan and player? Is there a difference
between Tiger and my fourteen-year-old, Julia Grace, an
artist, natural-born actress, and soprano who sings her
way through life? Is there a difference between Tiger and
my wife, Rita, the most beautiful woman in the world,
my *corazón,* and the soul of my family?

The question of the differences between humans and
other animals is not as simple as it might first seem. It
fleshes out like this: Is there a difference evident at a biological level, and is there a difference at a spiritual level?

It's necessary for us to wrestle with this distinction.
There is a well-intentioned global movement afoot to
bring animal rights into the mainstream. Some thirteen
states in the United States have now enacted laws that
make it illegal to chain or tether a dog to a pole or tree. In

Switzerland, it's against the law for a goldfish to be kept by itself. And in Spain, many legal rights now have been extended to nonhuman primates. There's talk of making it possible for an animal to have a lawyer defend him or her in court, and there's discussion about making it illegal for animals to be bought and sold.

The movement goes much further than advocating humane treatment for animals or ending unsafe puppy mills. At its core, the movement seeks to overturn "species-ism"—the idea that there are differences between humans and animals.[3] Pets aren't pets, advocates say, and pets aren't property—pets are people, too.[4]

But are they? Are humans any different from other mammals? Or are humans—dare we say—special?

The Law of the Jungle

Charles Darwin, the naturalist and founder of evolutionary theory, offered his thoughts about evolution and social engineering. It is quite easy to grasp from this passage of Darwin's what happens if we fail to make a fundamental distinction between our shared biology while recognizing our uniqueness, too:

> With savages, the weak in body or mind are soon eliminated; and those that survive commonly exhibit a vigorous state of health. We civilised men, on the other hand, do our utmost to check the process of elimination; we build asylums for the imbecile, the maimed, and the sick; we institute

poor-laws; and our medical men exert their utmost skill to save the life of every one to the last moment. . . . Thus the weak members of civilised societies propagate their kind. No one who has attended to the breeding of domestic animals will doubt that this must be highly injurious to the race of man. It is surprising how soon a want of care, or care wrongly directed, leads to the degeneration of a domestic race; but excepting in the case of man himself, hardly any one is so ignorant as to allow his worst animals to breed.[5]

My purpose for including that quote is not to vilify Darwin. Rather, I espouse the need for a much more honest conversation about what we actually believe when we discuss arguments regarding evolutionary theory from the relative perspectives of science or religion. The implications of what we believe, left unexamined, have led us to social Darwinism and the concept of "survival of the fittest"—a very slippery slope.

In the scientific community, of which I'm a part, too many scholars don the mantle of evolutionist as a holy writ to prove that God does not exist. Laymen then read the writings of these scholars, perhaps taking a cursory glance at their academic credentials and being swayed by them, and then they buy into the same line of anti-God thinking. As a result, we find a sort of bifurcation today between scientists and the religious. This is particularly noticed between the scientific community and creationists who hold to intelligent design theory. These creationists are sometimes mocked by the scientific community—and vice versa. Both sides of the debate believe it's an either-or argument.

But it's not an either-or argument. Both camps are right. Evolution can coexist quite happily with the existence of God. These two seemingly contradictory beliefs are not contradictory at all, but paradoxical.

Our problem is that we can easily be swayed by one camp or another without any real understanding of these assertions or their implications about the origins of *Homo sapiens* (which means "beings that know") on everything from scientific research to social policy. Evolutionary theory is much more nuanced and granular than either side cares to acknowledge. And so is faith in God, for that matter. As I wrote earlier, by no means am I advocating "blind faith" or a faith that somehow leaves behind intelligence and research. To the contrary, I'm advocating faith based on evidence.

Since this chapter is about what makes us unique as humans, we need to understand our own biological and "metabiological" origins as well, how we fit into this complex and unfolding story of humanity and the universe. If we humans fall solely into the evolutionary camp and are seen only as animals, then we don't live up to our full potential. Even so, as the Darwin quote asserts, there are grave dangers with this line of thinking. Namely, if we don't fully understand what makes us special, then we risk the commodification of our own humanity. We must remember that there are exceptions to the laws of biological evolution, the creation of the human mind being exhibit number one.

From nature's perspective, the emergence of the human mind is a wrench in the evolutionary machine, an aberration. Ask yourself this: What advantage is it for highly advanced nonaquatic primates to have a mind at all? From a purely evolutionary and survivalist perspective, we humans would have been better served if we'd turned out as mindless automatons. There would be no war or conflict, no ideology to believe in or to disagree about. Survival possibilities would have surely increased without a mind.

Yet somehow the human mind slipped from the reins of natural selection. It unshackled itself from the grip of biological causality, the chains of physical determinism. So we must conclude one of two things: Either evolution has a goal, which implies agency, or evolution is directionless. And even if we say that there is no self-organizing principle, no agency or direction, other than natural selection, then we still must contend that the creation of mind, the power and will to create de novo, its miraculous emergence and agency, is synonymous with purpose. The existence of mind reflects the unique human ability to create, without precedent or fixed laws. This universe-comprehending creation of consciousness is our minds' own work of art. With this gift we have the freedom to evolve, and transform ourselves beyond our material existence.

The phenomenon of human language is central to understanding humanity's place in the universe, and throughout this chapter we'll be discussing why language is the great defining watershed that changes everything. Neuroscience has helped us understand how our brains have evolved with greater complexity and utility through language to provide us with the capacity to think and to reason. It is the uniquely human brain that enables us to think abstractly, to create tools, and even to control our own destiny. The emergence of language represents a unique evolutionary algorithm, our mind's ability to codify ideas and to create garments in which to dress our inner states and ways in which to share them with others.

Inclusive with our development of language, we've created imagery and extended metaphors. We've developed an immensely complex and sophisticated system for one person's mind to connect to another person's mind—or even to a whole host of other people's minds, in the sense of mass communication. We don't see animals doing this. Animals don't have TED Talks. Dolphins or whales don't hold assemblies where they engage in political or religious debate.

Language is supreme. It is a gift unlike any other. Yet language is also a double-edged sword in the sense that words can be and have been used both to enlighten us and to inflame us. With language comes immense responsibility. Thus we need to fully understand the power of language. We literally can create and destroy the world within and the world beyond ourselves with the words we choose. The capacity for expression of thought is fraught with its own translational errors and risk of creating a "Tower of Babel" of disconnected space between us.

DNA: Decoding the Mind of God

Have you ever considered how language is woven through us? We don't simply *use* language. We *are* language. There is a law written by an anonymous author that all living things abide by. This law provides the constitution of life itself. We can conceive of this law as an a priori system of knowledge that gives nature its reason for action. DNA is the will of biology; its seeds transform the potentiality of life into its tangible expression, the blueprint that gives the butterfly its wings to fly.

Inscribed within the text of DNA are all of life's hidden messages and future possibilities, the intrinsic but unborn potential and tension between the "what is" and "what should be" of existence. Genes are the letters of DNA; their arrangement holds the ideas of the possible and emergent. Like any language, meanings can change through varied sequences. The biological discourse of gene expression has its own examples of editing, divergent meanings, and censorship—its variegated manifestations are dependent upon the faithfulness of expression from within the text of DNA.

As one example, there are regions of DNA that can be skipped or silenced, with corresponding changes in the makeup of the biological machinery. Changes in the letters of DNA are what give life its interpretative capacity, various modes of transmission that can change meaning. Nature has given its inheritance a repository of freedom, an ability to create a dialogue to respond and to adapt to what would otherwise be immutable biological constraints. In other words, with no pun intended, DNA is a language that evolves.

Genes are nestled on long chains of DNA called chromosomes. Chromosomes are so long that they must coil like a snake. These snakes are wrapped in layers of proteins, called histones, that control the activity of DNA and its expression. Like an ancient mummy, DNA can become uncoiled, which subsequently activates the hidden code of genes. As a result, the DNA itself does not change, but the expression of the genome can be altered. This process is known as *epigenetics,* a word that refers to modifications in the function of genes where there is no change or "mutation" in the "written" text or code of the gene itself.

The idea that living things adapt based upon experience and through the mechanism of epigenetics is also an underlying tenet of neuroplasticity, the idea that brains can change themselves based upon environmental influences. These transgenerational forms of inheritance can be traced through the evolution of language.

And what is the connection between epigenetics—the interpretive capacity of DNA—and the evolution of human language? It is the uniquely human possibility to transcend what appears fixed and immutable into alternative realities of meaning and being.

Spiritual Evolution

The creation of language was a quantum leap in evolution, the brain's own version of next-generation sequencing enabling it to manufacture its own reality of life. With our minds we have the ability to construct alternative meanings through the transcendent potential of human language. This transcendence in the development of language entails a progression in the complexity of representations, a conceptualization of knowledge that is not unlike how DNA encodes the inherent intelligence of life. Our ability to understand the evolution of DNA and ourselves, even to harness these evolutions as a means for us to understand ourselves and each other, is no less important for our survival as well. Language enables us to convey our essential essence and thereby understand each other more deeply. Minds change through language, as does culture and our deepest held beliefs.[6]

An example of how our minds have changed through time can be discovered through the evolution of writing. In the earliest recorded language, such as in the Phoenician and Hebrew alphabets, human history was chronicled in a certain direction and form. Reading and writing of those alphabets was from right to left. The particular syntax of language in this direction reflects a stronger dependence on right-brain activity for the initiation and propagation of ideas. As I discussed in an earlier chapter, the right side of our brain is more inclined to perceive reality in a more implicit and relationship-oriented manner. Concepts such as faith and belief are easier to comprehend when the right side of our brain is engaged.[7]

It was only much later in history that the Greek civilization changed the direction of writing from left to right. The Greeks were very much rationalists. They fostered logic and science, which became the cornerstones of Western civilization. They rejected many of the religious myths of more ancient civilizations, and a struggle between the believers and the rationalists emerged. The evolution of language and its effect on human consciousness was not a bystander in this debate, but actively shaped the seeming dichotomy between rationalism and faith, reflecting the brain's two dueling modes of perception. Nowadays, thanks to the directional decision made by the Greeks, we have a more difficult time comprehending faith and belief.

This struggle within ourselves has often been externalized. The perils of our inherent brain duality have been played out to such an extent that even those who proclaim strong belief and faith in God (and participate strongly in their respective religions) remain profoundly divided. These divisions of religion are seen not only in people of one faith setting themselves up against others with vastly different beliefs, but even among peoples of one faith: disagreements in doctrine have caused endless feuding and even bloodshed. As the European poet Rilke wrote, "Our mind is split. And at the shadow crossing / of heart-roads, there is no temple for Apollo."[8]

The difference between science and faith, as reflected in the use of language, is fundamentally a "disagreement over the existence of meaning."[9] It is a uniquely human characteristic and burden of the evolution of the human brain to freely explore value, purpose, and meaning, which we do through the use and interpretation of language. Through language we generate representations, models of reality we construct and generally agree upon. It is through these agreements that we can cooperate and live peacefully together.[10]

But we will always reach a limitation in our understanding through the use of these tools. The brain, in its evolutionary dash to optimize its representation of the world, paradoxically suffers at least a minor, and more likely major, distancing effect of language in interpreting experience. There is the experience of love and the words we use to describe it. There's a difference between images and ideas of the world around us and the words we choose to order those models for ourselves and each other. The limitations of language include significantly lost access to the world beyond words that poets have nobly and assiduously attempted to translate. If there isn't a word for something, then it must not exist, we erroneously conclude. The appalling sacrifice—what we forfeit—for the development of the expedient conceit of our lexicon of expression, however, is nothing less than a black hole between the "barbaric yawp"[11] of human experience and the symbolic representations of our imagination, even though we know that they are "untranslatable."

The defining characteristic of the mystical, religious, and wondrous encounter is its sheer ineffability. What is "worship" except our feeble attempt to express what cannot be expressed? Most people have experienced at least a few moments in life when words have failed them. We stand at the edge of the Grand Canyon and view vast expanses of multicolored layers of rock. We cannot communicate our wonder. All we can do is gasp. Or, still in the delivery room, we marvel at the newborn babe in our arms. We are grateful. We are staggered. But no words can come. We can only weep. Surely we are tapping into the intangible then. We have come to the limits of our language and have experienced something beyond language.

In doing so, our minds have brushed against God. We have encountered, however limited, the echoes of another world.

Moses himself confessed to a speech impediment and tried to turn down the job of communicating "the Word." He was, we are told, "heavy of mouth and heavy of tongue."[12]

The Loss of Language Is a Divided Brain

A neurological condition of language loss called aphasia demonstrates the power of language and the gravity of its erasure. Aphasia is not uncommon after strokes, but it's also rarely seen in children. There is, however, a rare condition known as Landau-Kleffner syndrome (LKS)—or, more accurately, acquired epileptic aphasia (AEA)—in which an intense immune attack occurs, primarily directed to regions of the brain necessary for language acquisition, comprehension, and employment. LKS can cruelly strip a developing child of language skills. Parents (and physicians) frequently confuse and misdiagnose this syndrome as something on the autism spectrum. MRI scans and electroencephalograph (EEG) brainwave studies are often helpful in making this diagnosis.

Researchers first described LKS in 1957. The disorder is characterized by gradual or sometimes rapid loss of language in a previously normal child owing to an anomalous immune-based trigger. It is treatable—though not always effectively—with immunosuppressive drugs.

Hanna, a nineteen-year-old girl living with her parents, experienced such a problem at age two. As a baby and young toddler, she exhibited fairly advanced development. To her parents' delight, Hanna reached the ability to speak in complex sentences at a very early age. But soon after her second birthday, she developed high fevers and often cried inconsolably. It was during the first few of these crises that Hanna first suffered significant regression of her verbal skills. She became abnormally hyperactive and walked with a bizarre and lurching gait. Often she displayed unprovoked tantrums and crying jags. These symptoms would slowly abate, only to re-emerge after some other triggering event, during which she would periodically "re-lose" her language abilities.

I asked the family to record such episodes on their video camera (this was in the early 1990s, before the advent of cell phones), and the documentary evidence they provided revealed a girl whose body became extremely rigid, whose eyes slowly drifted upward as if she were auditioning for a role as a zombie in some low-budget production. We were able to obtain an EEG recording during one of these events that confirmed there was abnormal electrical activity in the brain localized to the speech centers.

What Hanna was experiencing was an inability to make sense of reality. Without the predictive luxury of language, there were no patterns, only a threatening noise. The dissolution of the foundations of her brain's language centers meant that reality itself became deconstructed. Her brain directly and unabatedly experienced an inability to discern meaning, to harness order from the disease of entropy.

My interaction with Hanna over the years pointed me again to the absolute wonder and necessity of language.

Her story showed how, when language is removed from a person's perception of reality, reality looks totally different—and it's a much more brutal reality.

In some respects, we have lost our appreciation for language. We all suffer to some degree a spiritual type of aphasia. We're at a unique time in history, one in which we can see how the words of humanity can be twisted, misinterpreted, and altogether not received. We see the consequences of this in our collective social well-being. When we write and speak today, we tend to communicate as if we're detectives or reporters at a news station. Just the facts, thank you, ma'am. And everywhere today we see corollaries of this modern scientific thinking, extremely intense battles for the "right" to claim truth. There is little give-and-take. No wish to relax and understand the middle ground. People commit so strongly in their exclusive belief systems that they even are willing to kill other people and themselves to make statements that somehow prove the truth of their interpretation of reality.

How dangerous our rational, explicit brain-dominated thinking has become, how tragic are the consequences when the value of ideology triumphs over the value of life itself. Instead of our understanding how language and ideas can connect us, we are using the ideology we have created to divide us instead.

But surely there is hope. Surely we can reclaim the sense of immense responsibility we have with our language and use this power altruistically. To heal and not hurt. To unite and not divide. The answer may lie in an unlikely place, in what I call the "sacred evolution" of our language, in our story unfolding in the Mind of God.

Our Conversations with God

Through language, we hold the levers of creation. By virtue of being able to decode the symbolic we become aware of the fundamental tools of creation—a window into the machinations of the world and the Mind of God. We have the choice of using this gift in a responsible way, or we can ignore the potential power of words to alter our existence. All of the emerging gene-editing technologies are revealing the power of linguistic-based attribution—that all levels of reality are constructed in some symbolic form or another. Letters in both the biological and spiritual realms of our existence can transform, alter, and change the fate of life like no other tool.

So our lesson is this: Along with being blessed with this tool of language comes massive responsibility. Language gives us understanding; it gives us the key to alter our existence and to make choices about the existence of others. We have a much greater degree of accountability because of our awareness as such. There is hope. Certainly language can divide us as humans. But language can draw us together, too.

John Milton expressed this responsibility in *Paradise Lost*:

> *Immediate are the Acts of God, more swift*
> *Than time or motion, but to human ears*
> *Cannot without process of speech be told,*
> *So told as earthly notion can receive.*[13]

Indeed, the ability to create and reinterpret our language is also the ability to rediscover our godly nature. Milton reminds us:

The mind is its own place, and in itself
Can make a Heav'n of Hell, a Hell of Heav'n.[14]

Milton's poem alludes to the great potentiality of human language. It points to our ability through language to co-create our reality. Through the evolution of language and the expansive breadth of speech, we can evolve to something higher than merely our biological processes. We can discover our immaterial and transcendent godly souls. As co-creators of our lives, we get to write our stories in a manner that reflects altruistically on others.

Neuroscientists refer to the ability to construct our own personal narrative as autonoetic knowing, the brain's ability to create its own stories. Through language we create the meaning of our existence, but it also offers the opportunity to have communion, to know the "other" deeply as well. The biblical phrase *to know* can be read to mean a special union, a bond built upon shared dialogue and understanding. The unfolding of our lives can be conceived as a dialogue between God and his creation. We all encounter conflict along the way. In fact, every interesting story, whether fictional or journalistic, requires conflict at its core. Without barriers, challenges and challengers, and assorted other antagonists, we can't really find meaning or truth. As Martin Buber wrote, "The world is given to the human beings who perceive it, and the life of man is itself a giving and receiving."[15] These stories are conveyed to us and we transmit them through the garment of language. We are called upon through language and respond to it in kind. Buber continues, "Thus, the whole history of the world—the hidden, real world history—is a dialogue between God and his creature."[16] And it's through this dialogue that we may discover that love is the language of God.

What's the Meaning of Life?

God is the question to which our lives are an answer.

—RABBI JONATHAN SACKS[1]

Ultimately, man should not ask what the meaning of his life is, but rather he must recognize that it is *he* who is asked.

—VIKTOR FRANKL[2]

In Douglas Adams's classic sci-fi series, *The Hitchhiker's Guide to the Galaxy,* a huge supercomputer named Deep Thought is fed "the ultimate question of life, the universe, and everything." Seven and a half billion years later, when the computer has finally chewed on the question long enough, it spits out the answer:

"Forty-two."

It's a nonsensical answer, of course, although fans of Adams's work have sought for decades to find deeper meanings within it. Adams himself declared, "It was a joke. [The answer] had to be a number, an ordinary, smallish number, and I chose that one. I sat on my desk, stared into the garden and thought '42 will do.' I typed it out. End of story."[3]

In all seriousness, everybody wonders about the ultimate question of life. Meaning. Purpose. We want to know if there's some greater reason for our being born. Are we merely products of biology? Is there no greater meaning to life other than reproduction and survival? Is purpose only a figment of our imagination—something we humans made up? Or does life carry with it a marker of something greater, a mission beyond ourselves, something pointing to God?

An additional question dovetails with this. We find purpose not only in the personal realm but also in the domains of cosmology, physics, chemistry, and biology. Since all of life is a balance between creation and destruction, order and chaos,[4] patterns and sheer randomness, we're really asking this question as well: What does *purpose* mean? Purpose implies direction and intent, even destructive intent. So we need to be very careful with our handling of purpose. We need to see how it can shape both our biological and our spiritual destinies. Not only do our brains need purpose; they can't survive without it.

Thrown into the mix of understanding purpose is seeing its opposite force in nature. In science, *chaos* refers to that which is unpredictable and erratic. In physics—the study of motion—chaos is recognized as the second law of nature, and the concept of *entropy* (an irreversible or ruinous process) refers to the degree of disorganization among the components of a system. (Thanks to chaos and entropy, you can't unscramble an egg.) In biology, *chaos* implies an uncoupling of relations in an aggregate that normally are interdependent. When a certain organ or system becomes chaotic, its behavior becomes random and loses its order. For instance, in cancer, biological chaos leads to unchecked cell growth and metastasis, in heart disease the risk of sudden cardiac death. In the brain, serious conditions such as epilepsy and schizophrenia are associated with chaos and are aptly called "disorders." Serious pathology emerges when populations of cells become uncoupled and no longer receive or act upon other cells to coordinate their respective activity. Biological chaos within the brain can be episodic or chronic and even life-threatening. There is also psychological chaos that is manifested in relationships and globally; there are societies tilting into disorder as well.

The inquiry into the meaning of life is as "human" a question as there is, and we can try to understand meaning by understanding its relationship to the dynamics of order and disorder. The religious and philosophical traditions are steeped in this exploration; in both Greek and Hebrew theology, *chaos* implies a gap, a yawning—what's anomalous, unstable, and uncertain. In every facet of human life, the goal is to transform this state, to regenerate stability, to move away from this constituent element of experience.

We inquire about meaning—seeking evidence of an order—through our engagement with our minds, via thoughts, language, prayer, study, reading, and conversations with others. As we'll see in this chapter, we begin to find purpose through acts that bring order into our lives, through relationships based on love and compassion. The more connected we are to each other, the more purpose we find.

Neuroscience is particularly able to inform us about purpose. Our brains are specifically designed to create it. Life is dependent on counterbalancing entropy with purpose and order. The concepts we are describing are not merely metaphorical; every cell in our body has purpose, too. This purpose is to find balance—to maintain order in the midst of opposing challenges that will otherwise mean succumbing to disease. I experienced this firsthand when I was asked to see an adopted child with bizarre and extremely dangerous behavior. At no time in my clinical practice did I experience a more profound example of the psychological consequences of chaos—how severe emotional trauma and neglect have not only indelible behavioral consequences but biological ones as well—than when I met Dorin.

The Fear and Pain Channel

In the dark days after the Romanian Revolution that occurred in 1989, Dorin lived in a makeshift orphanage that teemed with five hundred other parentless urchins. The orphanage operated by the skin of its teeth and with an extremely limited staff. When Dorin's new adoptive father, Andy, came to pick him up, Andy was struck first by the stench of the place. Everywhere toddlers were crying, uncared for. Some wore dirty diapers—and looked to have been in that state for some time. Some toddlers simply stood stock-still, frozen in apathy like dolls.

But Andy and his wife, Janine, had committed

themselves deeply to Dorin from the first time they'd seen a faded Polaroid of his sallow face and deep-sunken eyes. At the orphanage, Andy filled out surprisingly little paperwork, and the next day he was flying back over the ocean with four-and-a-half-year-old Dorin in tow. The child didn't utter a peep, and he didn't sleep on the plane except in brief and restless shifts.

Dorin did not respond to Janine's wide smile when she met them at the airport. He didn't even glance at all the balloons she'd brought (thank goodness they'd already agreed not to bring Dorin's new grandparents or sister so as not to overwhelm him). Janine chided Andy for not cleaning Dorin up more. But more troubling than his filthy face and body was his size. He stood, emaciated and ungainly in the arrivals area outside immigration, looking more like a three-year-old than a boy of almost five. Janine couldn't help but think of those old films of Holocaust survivors, but she shook that image out of her head.

As soon as they got home, they tried to feed the boy, but Dorin wouldn't eat. Andy and Janine took him upstairs for a bath. That's when they first realized the matrons had squeezed his feet into shoes several sizes too small. More disturbingly, they noticed bite marks all over his arms. Were they his? Or were they from one of the other children?

The next day, Dorin's new big sister, eight-year-old Emily, returned from her overnight with her grandparents. With shrill giggles, she tried to hug her new brother. He responded by shrieking an unholy cry—the first sound he'd made. He scratched her face and bit her on the shoulder.

"My God," Janine said to her sister over the phone. "What have we gotten ourselves into?"

Neither Janine nor Andy ever considered giving up. But for months things continued in a violent and unpredictable vein. Dorin settled somewhat into his new surroundings but remained physically impulsive—sometimes to a dangerous degree. After days or weeks of utter sullenness, he might explode at some perceived slight such as a piece of skin left on his apple. He punched holes in walls and ran full speed into furniture. He smeared feces on the bathroom mirror. At the dinner table, to punctuate a lull in the conversation, he hurled knives at his parents and sister. He suffered frequent night terrors and sometimes screamed for hours despite Janine's efforts at gentle calming.

Slowly Dorin picked up scraps of English from his adoptive parents and sister. Had he ever spoken a word of his native tongue? Had anyone ever spoken to him? It was, Janine thought, as though they'd found a feral child from a primeval forest. When at last they could communicate with Dorin about things, Dorin had no conscious memory of his night disturbances the next morning whenever his parents tried to broach the subject. Similarly, he couldn't recall his violent waking outbursts, just hours after they occurred. Months passed like this, with a mysterious waxing and waning of abnormal behavior that had become cyclical and somewhat predictable.

One night when he was only seven, Dorin tried to set

fire to the house with a barbecue lighter. Andy, through some miracle, stopped him in the act. "One day he'll kill you all," Janine's sister said. And they feared she might be right. They knew there were such things as clinical sociopaths and psychopaths. Had they adopted one through no fault of their own? Janine was convinced that was impossible. She had always believed that nurture could trump nature. Enough passion could move any mountain; enough love could melt any iceberg.

Various professionals postulated that Dorin's episodes were symptomatic of a severe form of an emotional deprivation disorder. Then a new treating psychiatrist requested a neurological evaluation to rule out physical causes such as epilepsy—and that's where I came in. Dorin's was one of the strangest cases I'd ever seen. I thought about how his behavior was a result of a twisted form of survival of the fittest, a reflection of how his brain had adapted to the brutality of his earliest childhood experiences.

Darwinian evolutionary theory has been used to explain not only evolution, but most recently brain development and plasticity. Nobel Prize winner Gerald Edelman published a landmark book entitled *Neural Darwinism: The Theory of Neuronal Group Selection,* which advocated the idea that complex adaptations in our brains arise through a process similar to natural selection.[5] These circuits and patterns of activity in the brain can replicate in a survival-of-the-fittest way. Experience shapes our brains by affixing either life-affirming or life-defying messages in them. In scientific terms, changes in the connectivity between synapses lead to changes in the brain in a way that is similar to how evolution acts upon organisms. Our brains work best when they are connected. Or, put more simply, as the brain develops, it either selects only the best or sloughs off the worst.

We generally associate adaptability with positive forces. For instance, a child who suffers a stroke in infancy may quickly adapt and compensate by using other brain circuits to such an extent that undamaged areas of the brain assume the role of injured ones. No residual effects of the stroke are clinically visible. Indeed, the primary function of brain plasticity is that it is generally behaviorally useful. Neural circuits are created and reshaped to respond in ways that promote survival.

But the brain can be negatively affected, too. We can think of the brain as a muscle that needs to be strengthened for its primary job of interacting with other brains. Starting on day one, if you don't use it, you lose it. For decades, we've learned to "enrich" the experience of the animals we pen in zoo paddocks so their brains don't wither. Without stimulation from other animals, zoo animals (and other organisms) don't just simply grow bored—they withdraw, or they become violent (among other abnormal, antisocial, unproductive responses). Or both. Then they die.

Was Dorin's unusual behavior a kind of survival-of-the-fittest response, but in the worst way imaginable? We proceeded with a brain MRI study, which showed irregular brain myelin formation. The brain depends on myelin to effectively transmit messages from one region to another. When myelin, the white matter in the brain, evolutionarily developed in mammals, it produced a quantum leap in the function of our cognitive and emotional abilities by ensuring a high degree of connectivity between various brain regions. In human development, enriched and caring environments amplify the brain's ability to develop optimally, in large part through more vigorous myelin formation. It's really all about connectivity.

Some researchers have theorized that severe childhood neglect or abuse can reduce the brain's level of myelin.[6] Without emotional connections, physical ones become absent, too. One Harvard study has suggested that childhood neglect (abandonment and also sexual abuse) results in observable changes in the development of myelin. So what had happened to Dorin's myelin? What had changed his brain? The answer began in the cramped orphanage, the most inopportune environment for a brain at its most vulnerable period of development. I became convinced that what Janine called Dorin's "strange brain" had its roots in the atrocious orphanage where Dorin had spent his formative years. The sheer intensity of the qualitative emotional state had become hardwired into his nervous system, a network of synaptic activity operating as a 24/7 fear-and-pain channel. Unfortunately, we don't have a "reboot" function. We can't wipe our hard drive and erase these early childhood traumatic memories.

My colleagues and I prescribed and helped create a rehabilitation program for Dorin that included traditional psychotherapy, play therapy, movement therapy, and a combination of nutritional and medical support. We hoped Dorin could literally "think" his way into a healed brain. Time passed, and the intensive therapy progressed with positive results. The results were a demonstrable reduction of the frequency and intensity of Dorin's dangerous and antisocial behaviors. But even today, several years later, all this therapy has not fully "cured" him. And unfortunately we don't know whether the structural and functional changes that occurred in Dorin's little brain will ever be fully reversible.

It isn't difficult (but it is tragically frightening) to imagine that perhaps millions of children have endured horrific childhood experiences as traumatic as Dorin's. Children all around the globe are neglected and malnourished and exposed to unimaginably horrific violence and hostility. If we were able to examine the developing brains of these children as closely as we did Dorin's, we would likely find evidence that such traumatic psychological experiences have literally shaped and reshaped their developing brains.

The Biology of Belief

As we discussed earlier, concepts regarding order and purpose, or the opposite, chaos and meaninglessness, are not only metaphorical. The genius of Freud was that he implicitly understood that our metaphors, and even our beliefs, have biological roots. He postulated that there exists a connection between biological drives and extreme forms of abnormal behavior, such as what we had witnessed in Dorin's case. As an extension of his better-known libido concept, Freud posited what he termed a "death drive," an internally driven force of self-destruction, sometimes referred to as *Thanatos*.[7] Freud drew these conclusions from an ancient philosopher, Empedocles, who considered events in the life of the universe and in the spiritual life as governed by the two opposing, dualistic principles Love and Strife. These forces were elements of the universe at large and had an equal effect at the level of each human too.

While it is hard to understand how our brain would have a death drive, the validation of this theory in science led researchers Sydney Brenner, H. Robert Horvitz, and John E. Sulston to be awarded the 2002 Nobel Prize in Medicine[8] for their discovery of an embedded genetic code of programmed cell death, referred to as apoptosis. Apoptosis is a cell-death program inherent in every cell in our bodies, including our brains. From the very beginning of life, the disintegrating elements of this death drive are paradoxically and immanently active. Biologically, DNA contains within it the core messages of pre-programmed self-destructive tendencies.

Yet apoptosis, or cell death, is not all bad. Sometimes the "suicide" of specific cells is actually a good thing. For instance, in the womb, all of our toes are joined together at the start. If certain cells didn't die and disappear, then we'd all be born with webbed feet.[9] At its best, apoptosis is an integral part in the process of the brain and body's development and coincides with the establishment of synaptic connections and healthy lives. In the developing brain, for instance, there exists a delicate balance between these tightly regulated suicidal cell death pathways and cell survival signals. Apoptosis shapes the brain by refining and sculpting the developing synapses. At its worst, apoptosis can be very harmful. Some have even theorized that autism is caused by excessive apoptosis, and in certain diseases such as Alzheimer's and Parkinson's, studies show links to too much apoptosis happening within a body.

Even in the field of psychiatry, severe emotional distress and hopelessness may trigger latent cell death pathways. Psychological wounds can leave a form of brain scar tissue—an actual biological mark that can be quantified in neurons. In a recent study, researchers looked closely at the brains of patients who were deceased and who had suffered from treatment-resistant depression, and these scientists were able to measure the degree of apoptosis in populations of brain cells. They discovered a clear connection between biological death messages and the presence of severe emotional turmoil.[10] The cells that are chosen to die are those that have failed to establish meaningful connections; their detachment, a form of miscarried relationship, causes them to lose their purpose. It's as if the brain concludes that life isn't worth living: ultimately, these harmful messages get communicated to the genetic and biological machinery in tandem, resulting in a disconnect in vital pathways.

Meaningful Life

Life can become exceedingly dangerous when we lack meaning. Human anxiety results from a fear of separation, when our brains are divorced from purpose and filled instead with nihilism. Even on a physical basis, there exists within us a death drive triggered by hopelessness, a slow form of death in life when we conclude that life holds no meaning whatsoever. Neuroscience confirms that our brains seek purpose through our relationships. Purpose is coded right into the smallest subatomic components of our very being. I'll repeat my point: *purpose* is not something we humans simply dreamed up—it's coded within our very DNA!

Thus the largest and last question that emerges is about our application of this knowledge: Since there is purpose to life, what do we do with that purpose? How can we even advance our own evolution, if you will? Evidence bears out that experiences, and particularly our experiences in relationships where we establish connections, shape the brain. The brain, in turn, shapes our experiences. The influence of our beliefs is quite literally a matter of life or death. So we must therefore carefully safeguard our connections.

Gandhi famously put it this way:

> *Your beliefs become your thoughts,*
> *Your thoughts become your words,*
> *Your words become your actions,*
> *Your actions become your habits,*
> *Your habits become your values,*
> *Your values become your destiny.*[11]

Gandhi's quote feels inspiring to us if it's seen in positive terms. If we believe we can affect the world in good ways, then we'll start talking about those good ways, living out those best practices, making a habit of purposeful living, then ultimately shaping the course of our lives in beneficial ways.

Undergirding Gandhi's quote is neuroscience, which

is now able to demonstrate a direct effect of brain cellular activity on our cognitive and emotional states. The
biology is fact: What a person thinks, he becomes. If a
person continually thinks harmful thoughts, if a person
immerses himself in unchecked and prolonged rage, or
isolation, or any state of disconnection, then he can literally trigger harmful apoptosis and kill off his brain cells.
Conversely, the more a person exposes his brain to positive material, the more positively he will act. A brain can
be damaged, and a brain can be helped. Empathy, relationships, being altruistic, showing compassion, and living a heartfelt existence all promote brain regeneration.

Finding the Mind of God Through Our Hearts

Neuroscience has confirmed that our feeling states are a
portal of perception equal to or even surpassing in some
instances the role that vision plays in our experience of
reality. They are inextricably linked to the mental models we construct about the world, how we see ourselves,
and how we perceive and judge others. The primacy of
emotion, in both matters of consciousness and of faith,
comes from the fact that almost all of our relationships
are based on them.

There are so many examples of the link between our
emotional states and our brain function that an emerging field, neurocardiology, has cropped up, devoted exclusively to the study of the relationship between the brain
and the heart. Negative emotions such as hostility and
stress have been well correlated with heart disease, and
even sudden cardiac death (SCD).[12] Normal heart-rate

variability results from the coordinated coupling of "os-
cillators," specific groups of cells that regulate the balance between chaos and order in the heart. When this
carefully calibrated group of cells is radicalized in some
fashion, the normal equilibrium between the opposing
forces is catapulted into a life-threatening swirl of violence.

When I was a neurology resident, a young woman
suffered a massive stroke that was so extensive that the
swelling in her brain resulted in a coma. Strokes are very
rare in young people, and in this particular case the cause
was a mystery we needed to uncover.

Through a series of diagnostic tests, it was revealed
that this young woman had extensive heart damage resulting from a poorly understood condition called Takotsubo cardiomyopathy, more commonly known as
broken-heart syndrome.[13] In broken-heart syndrome,
the common denominator is often a sharp spike in
adrenaline-type hormones after emotional trauma or
severe stress. Various clinical cases abound in the medical literature on broken-heart syndrome—a victim of an
armed robbery, the sudden loss of a loved one.

We weren't able to elicit such a history. Our broken-
heart patient was already on a ventilator when she was
transferred to the neurology service, but she had a tattoo
on her arm that read LOVE HURTS. Every time I was at her
bedside and read this tattoo it reminded me of what Pascal observed: "The heart has its reasons of which reason
knows nothing."[14]

Every morning I would go into the wards and draw

her labs. It was difficult to find a vein; she was quite over-weight. It went on like this day after day. I would examine her neurological status, which never changed. So I would adjust her ventilator settings and the feeding tube, and I would manage the prophylaxis to prevent further clot-ting. I had no idea if she was or wasn't aware of her sur-roundings. She did not elicit any response to the tests we performed on her except for withdrawal to stimuli on her non-paralyzed side. Her prognosis seemed grim.

Then one morning I entered her room and did the usual ritual. I apologized for taking her blood, told her I thought she was improving, mentioned that her mother had been in to see her, and prepared to move on to the next patient. But this time I noticed she was awake, or at least her eyes were open. I knew she was conscious. Her eyes followed me and she responded to a simple com-mand. She made a request and indicated to me that she wanted to write something, so I quickly handed her my clipboard, a pen, and some paper.

It was an agonizing exercise in communication as she scribbled one letter at a time:

I . . . want . . .

She must need something. Is she in pain? Do I need to explain why she is in the hospital?

to . . . take you. . . .

Take me? What is she thinking here?

. . . to Red Lobster.

She pointed to me, and back to herself, just to make sure her request was perfectly clear. What? Red Lobster? You want to take me to Red Lobster?

To be sure she was making her point, she pointed to herself and then to me to make her intent clear.

It was her way of saying she wanted to thank me for my medical help.

After a while we were able to remove her from the ventilator. I had done nothing medically heroic for her, but she would get her second chance at life. Yet as I've thought about this case in hindsight, what I had provided without any awareness on my part was the connection she needed to give her brain purpose in the most isolated experience one can imagine—a coma.

I will never know what exactly she was processing in her brain during those brief moments during which I made small talk with her. What I had regarded as mundane conversation in retrospect must have reassured her that she would survive and thrive beyond her immediate medical crisis. What I did learn, though, is that communication is never a one-way street, even when we believe that no one is hearing us when we speak. In my profession, we sometimes forget that patients in a coma are often capable of hearing us discuss them, even if they are incapable of responding.

We never fully know what's in other people's minds—their thoughts, dreams, and fears—but our relationships are a testament to some deeper, wordless understanding in which we share our experience of the world. These seemingly one-way interactions nurtured her during the several weeks that she was otherwise disconnected, and proved to be the best medicine I had for her, whether or not I realized this at the time. Sometimes it's the smallest gestures that matter most. It's not love that hurts; it's our disconnection from it that causes our suffering.

It doesn't take a psychiatrist to realize that we are living in extraordinarily stressful and uncertain times. There is tremendous suffering in every corner of the world, a growing existential angst, and there are very good reasons for us to lose faith. I don't mean to imply that you and I are specifically suicidal. But too many people live with shades of hopelessness, with degrees of angst. Because of this everyday hopelessness, we live lives of despair and darkness instead of lives of confidence, strength, empathy, and compassion. Too many people are on the slow walk to suicide. They believe they have no purpose, no meaning. They've drunk the Kool-Aid that says it's all random—our birth was random, our life is random, and our death will be random.

That's a big lie. Life is about order, not chaos, and order is filled with meaning. We must spend our resources pushing back against disorder so that we can continue to have meaningful lives. With meaning in our lives, we humans live more deliberately, more compassionately.

Life is a reality that dispels the fiction of meaninglessness; it is a reality to which we are called and to which we can respond. Ultimate meaning is discovered through our connectivity and love for each other, and through this we attach ourselves to God. We are relational beings, and our brain finds its purpose through a deep awareness that the search for oneself is found at the nexus of self and other. This includes knowing each other as humans, and also knowing a God of empathy and compassion whose very DNA is reflected in our lives. When we are conscious that this positive purpose we so long for is actually an echo of God's mind, then we are able to live more securely. We can say with confidence that our purpose is to be empathetic toward others, and we can find direction and calm in that purpose. It is through the essence of our connectivity that we can resist the extreme pull of universal chaos that is gripping us all.

We see how purpose and meaning are synonymous, how they are the fundamental determinants of our fate. They both suggest there's a recognizable order and coherence to life—there's a greater narrative that brings together the many fragments and threads of our biological and immaterial existence. This knowledge is vital to our existence, and here for us to discover. When we talk about hope or faith or memory or love or empathy or communication or compassion, they all can become the essential nutrients of our soul. They are nonmaterial, but they are no less essential than the medicines we routinely rely on.

The medieval Jewish philosopher Maimonides once wrote that the world hangs in the balance.[15] This balance is like a butterfly effect, where even a small action can ripple through the densest walls of resistance. With purpose and through purpose, we can effect change, tipping the balance toward holiness, an integrated existence. We humans do have a choice. We can act in the midst of these challenges. We can instill order and meaning through deliberate, benevolent relationships. We have the capacity to act, to lay the foundation of love in all our interactions, engagements, and moral decisions. This is the ultimate meaning of our lives; at the nexus of both internal and external opposing forces, we both discern and shape our destiny. We were created to transform the material world, to bring order to an otherwise seemingly random and capricious existence, and to shape our lives by instilling our godly nature in the acts we perform.

This is our purpose: to discern and create a deep unity between ourselves and our Creator.

6

Are We Free?

Man can do what he wills, but he cannot will what
he wills.

—ARTHUR SCHOPENHAUER[1]

We have been slaves in the outer world but free
men and women in our souls and spirits.

—MAHARAL OF PRAGUE[2]

All of our experiences in life—our actions, feelings, and
motivations—are based, in part, upon the activity of our
brains. Hippocrates noted as much in *The Sacred Disease*:
"Men ought to know that from the brain, and from the
brain only, arise our pleasures, joys, laughter and jests, as
well as our sorrows, pains, griefs and tears."[3]

Neuroscience is based on the principle that human
thought and behavior are firmly rooted in the brain's biological function, and on the strong clinical observations
between various brain injuries and subsequent signs and
symptoms of neurological diseases such as loss of language, paralysis, and blindness. This begs the question:
How much of what we do is actually volitional?

This question of predeterminism even has legal bearings in death penalty cases. For example, in 2015 Cecil Clayton was executed in Missouri after being convicted of murdering a police officer. But in the 1970s Clayton had suffered a severe brain injury that resulted in neurosurgeons having to remove one-fifth of his frontal lobes. Their normal operations are essential for judgment, impulse control, and even matters related to moral behavior. Lawyers for Clayton argued as much, but the U.S. Supreme Court rejected the idea that this type of brain injury could have led him to commit murder. Such extreme examples in criminal law lead us to ask: How much of our brain's anatomy determines our destiny, and are we therefore not accountable for our actions?

One of the most unusual cases neurologists encounter is a condition called alien hand syndrome. Patients with this condition have hands that appear to be locked in a tug of war, where one hand may take objects away from the other hand or have one limb trying to restrain the other. When people observe these unusual cases, it almost appears that each separate half of the brain has its own private experience—desire and control over a person's behavior. In other words, who is really in control? Who is deciding for us? Is there a single *I* or are there multiple loci of control?

These questions are what we want to examine more fully in this chapter. Part of the study involves learning what true freedom means, and part of the study involves backing up a step and examining if actual freedom exists for us at all. It's the age-old debate between free will and predeterminism.

Are our lives fated for us? Have the gods—or has our

own biology—handed to us an unalterable template such that we must live the life predetermined for us to live?

Or are there options open to us? Is there such a thing as self-directed actions, whereby we choose and even create the very thoughts and movements of our biology? Are we able to live a life whereby we create our own destiny?

Are we free? Or are we not free?

Or are we, at the risk of sounding too middling, a bit of both?

Which Came First, Our Essence or Our Existence?

To understand the overall concept of predeterminism versus free will, we must ask ourselves, Which came first, our essence or our existence? Another word for *essence* would be *consciousness,* our "self-ness," or soul. Another word for *existence* would be *biology,* or our physical presence.

Let's consider the pure determinism viewpoint first, which would claim that our essence came before our existence.

Consider any mass-produced manufactured object such as a tennis racket. Before the engineers start the manufacturing process, they decide on the purpose a tennis racket will serve. They envision two tennis players on a tennis court who need rackets in order to play. A purpose is born. Then the engineers begin to create the actual rackets—the "existence"—and they do this by creating a prototype—one original tennis racket that acts as a model or standard for all tennis rackets to come.

To make this prototype, engineers take wood or steel

or fiberglass and shape it into the familiar tennis-racket shape. They experiment with different types of cord and filaments and string the head of the racket. They wrap the handle of the racket in grip tape—and voilà, an entire tennis racket has been conceived, designed, and created, and is now ready for production.

As part of the process of creating the prototype, the engineers also make a blueprint for all future tennis rackets. When the tennis racket company needs to produce a large quantity of tennis rackets to sell to consumers, they don't try to come up with an original design for each subsequent tennis racket they make. They simply look at the blueprint and work from there. Once that blueprint is drawn, any tennis racket made with that blueprint will always be a tennis racket. The blueprint combined with the raw materials won't produce a chaise longue or a bathtub or a rocket ship instead. A blueprint for a tennis racket will always yield a tennis racket.

Likewise, if something is originally built to be a tennis racket, then it's always going to be used for its intended purpose. Certainly in a person's imagination a tennis racket can be used for something else—an electric guitar, perhaps, in a lip sync competition. But most definitely it will always be a tennis racket intended for use by tennis players on a tennis court, never an actual electric guitar.

It's logical to see that the reason (or purpose) for something's existence should come first—and then everything will flow from that. Specific to this study, the essence of humanity is that humans were purposed to be humans. There's a blueprint for humanity encoded in our DNA, and all humans are built according to this same basic blueprint. The concept of being human, in the Mind of God, is comparable to the concept of producing an object following the origination of an idea of what that object will ultimately be used for. Thus, each individual is the realization of a certain concept within the divine intelligence.

On the flip side, there's pure free will. With free will, secular humanism and existentialism have reversed this entire equation and asserted that our existence came first, then our essence followed. We are creators of our own existence, including our minds and our consciousness. As such, we can shape ourselves and our world in every possible way. No one designed humans—not a personal Judeo-Christian God, and not even an impersonal intent within biology. We appeared here on earth through evolution, and then our task became to shape ourselves.[4]

According to this thinking, again using the tennis-racket illustration, there was no reason for designing a tennis racket to begin with, no concept of tennis even. There were no engineers other than the components of the various pieces of the tennis rackets themselves. And there was no factory other than the factory designed by the components of the tennis rackets themselves. The tennis rackets are free to be tennis rackets, but they might change their purpose, too, and become something else, if there is enough inclination and time to do so. Since we humans exist first, without an essence, the only meaning is for us to create one—the only meaning we have is the meaning we create for ourselves. Thus the "naked ape" theory is born. No one designed humanity. No one handed it a purpose. We appeared here, and then we needed to shape ourselves.

But which viewpoint is correct? Predeterminism or free will? Are we chickens, or are we eggs—and can we ever understand which came first?

I contend we must hold to *both* predeterminism and free will. The magic of human evolution was that we started with a template; we started with intent and predeterminism. A divine spark created man from the dust of the earth, and then God breathed a soul into man. Our consciousness both has purpose and can create purpose—and that's what gives us our free will. I will discuss later in this chapter the biological origins of free will as a fundamental human attribute within our brains.

There are constraints, sure. It's not all free will, and it's not all predeterminism. Both are equally true, and if that sounds like a merger of contradictions to you, then you'd be right. We are predetermined, yet we are also free. We are predetermined in terms of our biology, mostly. But we are free in terms of our consciousness, mostly.

This requires a closer look. Such a merger of apparent opposite viewpoints doesn't actually need to be seen as a contradiction at all. Think of it instead as a paradox, and paradoxes exist in science all the time. Science seeks to understand life's mysteries through empiricism and reason, rationalism and logic. We find that those laws of nature that appear at first completely irrefutable are prone to cracks. These cracks have come from discoveries that are anything but orderly: relativity, non-locality, indeterminism, and the uncertainty principle have helped us grasp our enigmatic universe through paradox rather than through assurance.[5]

For instance, the light of the universe is now considered one of the greatest mysteries of nature. It can be at once a particle and a wave, two entirely different conformations. Thus, knowing that the existence of paradoxes is not unique to this study helps us relax. We can view the paradox of free will versus predetermination as more of a "puzzle" and less of a "contradiction."

Theologians of various denominations have long debated the question of free will versus predetermination, and many proponents hold to a paradox as the answer; free will and predeterminism are not at odds with each other. With this hybrid doctrine, God is definitely sovereign. He rules and reigns over the universe, and God can do whatever he wishes. He transcends creation. That's the predeterminism side of the equation. Yet mankind clearly has many choices to make, including the awareness by each person that what he does in his life not only directs his fate but affects the lives of countless others. That's free will. God didn't create robots; he wants people who are able to choose between right and wrong and also to love him or not without any compulsion. This hybrid view embraces the paradox rather than kicking against it.

Similarly, within scientific study, both predeterminism and free will can paradoxically hold weight. Evolution represents being created in God's image: our essence is truly free; we can and do choose how to define and even create reality. Our essence is no longer deterministic, and we are free to define our essence. But within our evolution, we were designed with intent in mind. This is not a contradiction but a puzzle.

Humanity became free because God created our essence as such. Existence and essence occurred in tandem in evolution; our nakedness was our freedom to do as we choose, to realize we are not constrained by anything external. The only limits we have are those that are self-imposed. We can choose not to believe in God. But if we do not believe in God, then we must also agree that we cannot blame God. If we did not have free choice, if we were pre-programmed automatons, then our existence would have no purpose or meaning.

The sad thing is that too often we misinterpret freedom to mean that there are no obligations of the spirit. The Hebrew word *chârath* can mean both "freedom" and "to engrave."[6] The idea is that the only thing predetermined or fixed about our existence is our freedom. There is a divinely given attribute of being human that is encoded in our DNA, an attribute that endows us with a level of awareness to experience God in the full freedom of the human mind and heart. But with this freedom comes an existential responsibility that binds us to the world. It does not release us to do whatever we feel like doing.

True freedom is a key to a seemingly locked door. We are all handed the key and invited to open the door. But it's our choice whether or not to progress. Once through the doorway, we find on the other side a wide-open expanse of possibility. When we embrace true freedom, we are free to live the lives we were meant to live. We can live lives of real purpose.

So we are free. Yet we are not free to do whatever we wish.

We are free to truly help the world.

And that is true freedom.

Testing Free Will in the Lab

Several years back, a major award was given to a neurologist named Dr. Benjamin Libet, who worked to prove that our consciousness is predetermined, and predetermined only.[7] Basically, his argument was that everything has a cause and effect, including our brains. If your hand feels hot, for instance, it's because you touched something like a hot pan on a stove. The heat gets passed along into the process of your brain's neurons. You feel the hot temperature, and a reaction of the brain is felt in your brain. So Libet's larger question was fleshed out like this: Does our sense of will have a predetermined origin? In terms of our choices, are we behind the eight ball or in front of it? Is everything we do predetermined by an identifiable cause?

To give Libet credit, his hypothesis contained a lot of sound reasoning—and I agree with part of his conclusions. According to the law of causality (a law that has been understood and endorsed by minds such as those of Plato and Aristotle on forward), only one outcome can result from an initial existing state. If I have a ballpoint pen in my hand and I shake the pen, then I won't produce a tsunami. Why? Because the pen isn't big enough; it doesn't have enough mass and gravitational pull to produce that type of outcome. Only an adequate cause can produce a tsunami, such as the shaking of something as massive as the earth.

In such a deterministic universe based on cause and effect, we can regard reality as synonymous with predictability. "What does determinism profess?" asks philosopher William James. "It professes that . . . the future has no ambiguous possibilities bidden in its womb; the part we call the present is compatible with only one totality. Any other future complement than the one fixed from eternity is impossible."[8] Meaning, if I want to build a house, then I must be able to predict that I can build a house. I can only predict house building if I first have the intent to build a house. So there must be intent woven into our biology. Thanks to the law of cause and effect, there must be a designer before there is a template. And because of that template we must factor predeterminism into the overall equation. But is there free will, too? That's what Libet tried to disprove.

In one of Libet's experiments, for instance, people were told to press a button with one of their fingers. While they did so, Libet monitored their brain activity. He showed that we can actually physiologically measure through EEGs an impulse before an actual action takes place. In this case, the brain fired first, the person became aware of his decision to move his finger second, and then the actual finger moved third. There were gaps of hundreds of milliseconds between each of the three stages. Libet concluded that our brains appear to decide to move before we have any conscious intention to do so, suggesting that the conscious decision "I choose to move" is more of an afterthought than the causal determining force with a simple motor task. The implication is that we're not in control of our free will, because we don't have it in the first place. Before our actions take place, there is an unconscious signal that precedes each action. A person's brain commits to certain decisions before the person is aware of having made the decision.

Even Malcolm Gladwell has commented on this: "[Research suggests] that what we think of as free will is largely an illusion: much of the time, we are simply operating on automatic pilot, and the way we think and act—and *how well* we think and act on the spur of the moment—are a lot more susceptible to outside influences than we realize."9

But there were problems with Libet's conclusions, too, and they've been robustly debated for the past thirty years or so. New findings show that he timed the gaps wrong, for instance.10 Other studies show that his entire conclusion was flawed. For example, Dr. Angus Menuge, professor of philosophy at Concordia University, Wisconsin, writes, "If you look at Libet's experiments closely, there was a prior conscious decision by the instructed subject, then a readiness potential, then awareness of that readiness potential, and then a movement. So one can still say that a distal conscious decision was the cause of the movement, even if the proximal cause is the readiness potential."11 In other words, the will to move precedes both the decision to act and the act of movement itself.

And Dr. Massimo Pigliucci, professor of philosophy at CUNY–City College, writes, "Libet's experiments demonstrate that we make unconscious decisions ahead of becoming aware that we have made them. I doubt that anyone who has caught a falling object before realizing what he was doing would be surprised, and I doubt anyone would seriously take that sort of experience as evidence that consciousness doesn't enter into deliberative decision making."[12]

Libet's results have been reproduced and refined in numerous studies, including an experiment published in 2011 by Itzhak Fried, a professor of neurosurgery and psychiatry at the David Geffen School of Medicine at UCLA.[13] Fried replaced Libet's EEG recordings with electrodes that monitored single neurons and found that the readiness potential isn't just some nonspecific preparatory signal as some have argued, but rather brain activity that predicts both whether a person will move and what hand he will use before he makes those conscious decisions. Again, this seems to add weight to our subjective sense of free will in which our experience tells us that our conscious decision to move is what sets that decision in motion. In actuality, things are already set into motion long before any conscious awareness of that decision is made.

There is free will—our indivisible and divine essence—the essential force of life. And there is predeterminism—the manifested existence of cause and effect. This is not a contradiction but a mystery and paradox.

The Uniqueness of Consciousness
As mentioned, I'm a neurologist who "believes in belief," and I realize that requires some explanation along the

way. It is an empirical fact that our entire existence is based on some level of belief. The entire operating system of our brains—all our perceptions, sensations, thoughts, and emotions—are matters of belief. We are no doubt "psychologically primed" for religion. This belief in mind (in duality) is dismissed by some thinkers as an unreal by-product of our brains rather than an essential attribute.

To explain further—the classic theist believes in a transcendent being, one who is actively engaged in the inner workings of creation. The belief in purpose and agency is the central argument of a theist. Theists base their belief on an implicit understanding that we are endowed with a natural predisposition to seek and create value and intent. Our brains are able to inquire. They were indeed designed this way, just as the universe is perfectly suited for us because we are here—this, again, is the idea of the neuro-anthropic principle.

Yet why are we programmed this way? Atheist author Richard Dawkins critiques this idea: "We are biologically programmed to impute intentions to entities whose behaviour matters to us."[14] What evolutionary significance does this design of the human brain have if this is a false belief, an illusion? Why would we assign purpose and agency in the absence of any higher truth? Is the evolution-based belief in "intelligent design" a good fit for the survival of our genes? Or is it a profound detour in our quest for understanding whether there is any meaning to our existence?

To answer, we come back to the existential idea of whether existence precedes essence, or essence precedes existence. The humanists argue that first there is physical existence and second there is essence, but the latter is defined and created through our free will. I argue that there is first essence—God's essence, or the intent of biology. Through God's metaphorical mind both our existence and our essence were created. It's through our evolution that we rediscover this essence through our own consciousness, and by doing so we discover the essence of God as well. The Nobel laureate Sir John Eccles succinctly summarized his many years of studying the brain in his masterpiece *Evolution of the Brain: Creation of the Self*: "Biological evolution transcends itself in providing the material basis, the human brain, for self-conscious beings whose very nature is to seek for hope, to inquire for meaning, in the quest of love, truth and beauty."[15]

Central to a theist's belief system is what's called *irreducible complexity*—the concept that if you have a complex system of interacting parts and remove any one of the parts that make the system function, then the whole will cease functioning, too. For instance, if you want to catch mice with a mousetrap, you can't begin with only a spring and expect it to work. Similarly, you can't remove all the parts of the mousetrap except the spring and still hope that it will function correctly. The mousetrap will only function if all the parts are in place. For a complex biological and organic system such as our universe, planet Earth, and human life to function, each and every component must be in place. Thus a complex system is incapable of being reduced or diminished.

Dawkins sharply critiques irreducible complexity—and he's correct about a few things. It's scientifically proven that almost every biological thing in the universe and organic life can be broken down into components. You can be as much of a reductionist as you want, and you'll find a preexisting "something" that gave origins to the next thing. For instance, you can go to the atom, and there's even smaller things than that—quarks.

But there is one singular example in evolution that is irreducibly complex, and that is human consciousness. When we enter the realm of consciousness, we realize that our empirical sciences begin to lose their gravitas, the certainty of our perceived reality. Consciousness is functionally indivisible, and our conclusions about it can never be empirically tested. We know human consciousness is irreducibly complex. If we damage, remove, or repress a part of consciousness, then the whole ceases to function in a normal, integrated way. Indeed, the concept of irreducible complexity is central to the theme of our book—that our essence (human consciousness) is our highly specialized vehicle (a vehicle designed thanks to evolution) to comprehend the Mind of God. With sophisticated brain imaging studies of key regions within the human brain, we're certainly not seeing mere on-off switches. Instead, we can see on these maps, which can more aptly be described as networks of connectivity, quite literally, the soul within the machine.

When Free Will Is Lost

Neurologists sometimes act like detectives. We look for clues, we develop hypotheses, and we need to learn when to trust our instincts and when not to. When a fourteen-year-old boy named Ken ambled into my consultation room accompanied by his clearly apprehensive mother, I had no idea at the time that my medical workup would become a quest to locate the proverbial smoking gun of free will within the human brain.

I noticed that when Ken walked he exhibited an odd gait, dragging his left leg. This could have been a telltale sign of a stroke, not something you expect to see in a child. I wondered what could have caused the boy's weakness in that lower extremity, and I assumed that the referral by a lawyer meant I was going to hear about a malpractice case. It was not a malpractice case. Ken's mother, Sharon, provided most of the history for me while the boy stared blankly out the window. As a single parent, she was raising Ken and his younger sister, and she was feeling overwhelmed. Her husband had left her a year before what she described cryptically as "the incident with Ken at school." Reluctantly, and after some prodding, Sharon relayed that Ken was in legal trouble. He was facing criminal charges and the possibility of some serious time in juvenile detention. Her lawyer suggested a neurologist could help. But how?

Ken's legal bind had begun when another student reported some type of nonconsensual sexual encounter in the school bathroom. Ken allegedly molested a child three years younger than he was at the time.

While his mother related the agonizing details of the event as communicated to her by the school's administrators, I watched the boy's facial expression closely. There was no discernible reaction from him, not the slightest hint of any sentiment or emotion. By contrast, Sharon could barely hold back tears. She looked even a bit hysterical.

I asked the boy a few preliminary questions about the incident, but Ken neither denied nor tried to explain what had happened. It wasn't clear whether he didn't understand the indictment against him—or he just didn't care. It was like he was made of ice.

Psychopathology is defined as an inability to recognize, care about, or respond to the mental state of others.[16] Juveniles by nature can act truculent or indifferent. They often lack the communication skills to clearly articulate what's going on. So it's always tricky to diagnose a juvenile who might fit the clinical profile of a sociopath, a kind of "young monster" with an absent conscience. But the longer we talked, the more I began to wonder if Ken fit this bill. Anytime I asked him to explain his actions or even tell me what he was feeling, he just shrugged and answered, "I dunno." I got the impression we were sitting with a robot some human operator had unplugged.

"Ken, how do you think that other kid might feel about what you did?"

"I dunno."

"Do you think that person might be hurt now, or scared?"

"I dunno. Maybe."

"Are you aware how much this incident has been hurting your mother?"

Sharon now openly wept as she sat beside him.

"I dunno. I guess."

The boy's affect was completely flat. His voice was completely monotone. It wasn't mere teenage indifference. The lights were on, but nobody was home.

Hiding my deep concern, I asked the boy's mother how she thought I might be able to help. She said her pediatrician had referred Ken to a local psychiatrist. The psychiatrist concluded that the boy suffered from a learning disability and possibly ADHD, which was impairing his judgment. He based this conclusion on reports and observations of poor focus and low motivation at school: Ken had become increasingly impulsive and was failing most of his classes. However, a trial of several different stimulants—the typical drugs used to treat ADHD—had proved entirely ineffective. Ken was clearly a damaged child. But why?

Probing more, I learned some details of his medical history. Ken was born with a condition called congenital hydrocephalus, a buildup of excess cerebrospinal fluid in the brain. As an infant, he had had a shunt implanted in the right lateral ventricle of his brain to relieve the pressure. Two years ago he had undergone a shunt revision, normally a relatively routine procedure. But doctors had told Sharon that there had been some "complications" (lingo doctors commonly use to say something got botched up). In the procedure, the shunt had injured Ken's right motor cortex, leaving him with the residual weakness I noticed in his left leg.

My Sherlock Holmes mentality kicked in. I wondered whether there was any connection between this botched procedure and Ken's behavioral issues, although at first this seemed unlikely. Sharon noticed only that her son sometimes appeared "depressed," which she attributed to the stress of regular hospitalizations and follow-up visits with doctors.

I asked Ken outright if he ever felt depressed.

"No," he said. "I don't feel anything."

That was a clue. For the first time in the conversation, he hadn't answered with "I dunno."

But his mother hadn't caught the clue. "Ken," his mother said, her voice cracking, "you could go away. They could take you away to a jail for boys."

The boy gave no response. No nod, no flicker behind the eyes.

Ken had been telling the truth all along. He was not depressed. Depression, at least, is a feeling. But Ken did not feel anything. He was completely apathetic. The real Ken was gone.

Our session was over, at least for the time being. Because of the neurosurgical history, I referred the boy for a functional MRI (fMRI) and a diffusion tensor imaging (DTI) test, although I was hard-pressed to make a direct connection between my findings and his current anti-social behavior and legal troubles. DTI had just become available at the time: It could detect injury to the brain in cases where previous studies appeared normal.

A few days after these studies were completed, I received a direct call from the neuroradiologist, which was highly unusual. "Thanks for this one," he said. "I've never seen a case like this."

It turned out that the shunt placement had impaired not only Ken's motor cortex—the region of the cerebral cortex involved in the planning, control, and execution of voluntary movements—but also a significant amount of white matter in his right frontal lobe. White matter is the "subway" or "cable network" of the brain, the fast-moving information transport system located deep under the surface that connects each far-flung outpost of gray matter in the cerebrum to all the others. The damage caused by the revision of Ken's shunt indicated such a disconnect—and explained the reduced level of connectivity between reasoning and empathy.

The injury inflicted by the procedure accounted for his personality change and his lack of regard for other people's feelings. Ken wasn't born a sociopath—his brain damage simply made him act like one. The damage from the shunt placement prevented him from constructing any meaningful model he could use to understand what he did to his victims—or even why he did what he did. His power of reasoning was no longer connected to his empathy. He was quite simply disconnected.

Both the prosecuting attorney and the judge understood my testimony about the unique circumstances of Ken's case. He received needed medical and psychological intervention and was never incarcerated. Fortunately, over time Ken recovered, and gradually he became more and more like his old self, eventually showing genuinely deep regret and guilt for his actions—which of course required further treatment.

Ken's case shows the power of the brain anatomy to

direct our behavior, including our thoughts and feelings. The soul acting within our brains always has some level of biological constraint. A principle of evolution, and a corresponding principle of neuroscience, is the idea of causality—that everything must have a cause. The issue is whether our conscious states are a cause or a consequence of our brain's underpinnings. So really the issue of free will, psychologically speaking, is whether our brain states are directed from above the organ itself or whether our hardware—all of the machinations of our brain's inner machinery, the brain stem, and thalamus—drive our conscious states from the bottom up. So the other questions that we must grapple with are these: Where in our brains are we free? In what part of our brain's operational system do we find freedom of will?

The Real Freedom

The locus of identity, the anatomical center of our free will, is a region of the brain responsible for the authorship of the story we call ourselves. It's where my unique profile recalls that my name is Jay Lombard and that I trained as a neurologist and am now writing a book about neuroscience, faith, and the human soul. It's here in this particular group of brain circuits and pathways that I remember that I love my wife and children. And it is through these neurons' unfathomable basic processes that my most personal thoughts, feelings, hopes, and emotions are born and linger in my mind, both generating and recording my own personal history and not someone else's. The brain is a master storyteller, and it is within these anatomical borders that we weave and tell our stories.

We can regard this substrate as the source of our capacity to think and to imagine, the power and will to create, to transcend the limits of our existence, to choose for ourselves what we believe. We are *Homo sapiens*—"beings that know," born with the capacity of self-reflection and self-determination. To act with determinate purpose is to have an ever-abiding awareness that we are co-creators with God through our ability to create a shared version of reality. There is a tremendous responsibility that comes with this freedom. We never know what breed of butterfly will bloom from this cocoon; this universe is a Prometheus unbound.

Since our own minds and the minds and thoughts of other people are irreducible, and not directly observable, we have no direct way of verifying that other people even have minds at all; we can only ever implicitly infer the existence of minds in other people based upon our own experiences with our own minds. We therefore refer to the existence of mind as a "theory." As noted in chapter 2, the capacity to make inferences about the mental state of others is included in theory of mind. We know that others believe because we believe and that others feel because we feel. But what precisely does this have to do with our freedom?

Cambridge University psychopathy professor Simon Baron-Cohen has theorized that autism is a theory-of-mind impairment, an inability to fully and completely take someone else's perspective.[17] Anatomically, it has been proposed that the metarepresentation impairment of autism is also due to some reduction in functional connectivity and the computations involved in theory of mind leading to a state of "mind-blindness." Some may incorrectly assume that this state would render one incapable of knowing how someone else may feel.

But the human mind refuses to be bound by its roots, and anatomy does not automatically equal destiny. Naoki Higashida, a thirteen-year-old autistic Japanese boy, mastered the capacity—one keystroke at a time—to use written language to express his inner voice. Although he still cannot speak, he has written a bestselling book that has been translated into many languages. This autistic child, seemingly unable to grasp the true intent or feeling state of others, observed the following about those of us who purportedly have an intact theory of mind: "I think that people with autism are born outside the regime of civilization ... as a result of all the killings in the world and the selfish planet-wrecking that humanity has committed. . . . Autism has somehow arisen out of this. . . . We are more like travelers from the distant, distant past."[18]

Autistic people, seemingly disconnected from us, are like canaries in the coal mine. The word *autism* is derived from the Greek word meaning "self" and is used to describe behavior in people who appear cut off and disconnected from others. A common symptom of autism is extreme fear of strangers. But who is more fearful of others, the autistic child or the society she finds herself living in? Is it not fair to say that we are living in the Era of Autism, where alienation and distrust are ubiquitous and pervasive, and where we have progressively devalued our inner subjectivity? We have become "autistic"—the technologies that purportedly serve us have become the tools of our oppression instead. We experience a level of paranoia and distrust between us, replaced by a panopti-cal society and near absolute verbal surveillance. Is this freedom?

If we are truly free, what, then, is our responsibility? What lessons are there for us to comprehend and act upon with a brain designed for freedom—and what re-sponse is required from us in turn? In other words, if we are free, what are we free to do?

Free will means our ability to construct our own nar-rative about the meaning and purpose of our lives. We are free to believe in God, or to deny the existence of a Creator. We are free to act morally, or we may choose not to. Only humans are endowed with this potential, or the capacity, to go beyond the limits of our existence, to freely choose for ourselves a particular path. We can even choose to act inhumanely. We hold these truths to be self-evident: our brains seek their own freedom, to be autonomous, free to create their own destiny, unshackled from the chains of an absolute biological determinism. If we imagine the world on a scale that can be tipped in one direction or the other by our own actions—that our collective fate rests on the responsibility we have as individuals for each other—our first creative act of freedom must be forgive-ness. We believe that our lives are based exclusively upon cause and effect. We are pre-programmed to react: If we are hurt, it is normal to seek revenge or reconstitution. This propensity is based upon our deterministic nature, the conditional and expected. So much of human history is based on this principle, the expected response to an offense, but revenge gains no new ground.

There can be no free will without the ability to forgive. Researchers in Italy who have studied in depth the function of the regions of the brain involved in storytelling and identity have arrived at a startling finding consistent with a basic underpinning of theology. Emiliano Ricciardi and his colleagues at the University of Pisa examined the brain correlates of forgiveness using fMRI. They discovered that a key component of this region within the brain was activated only when individuals were able to experience empathy and thus forgive "others."[19] Free will means having the will to act unconditionally, the ability to transcend the constraints of causality and thereby live in the midst of the freedom and potentiality of God.

The process of forgiveness is the only true freedom. It endows us with the potential to act in a fully unconditional manner. So much of human history, both at the personal level and at the level of nations, is based on the principle of revenge, the expected response to an offense. But revenge gains no ground.

Forgiveness is what's unexpected. Hannah Arendt, who authored *The Human Condition*, wrote that forgiveness is a "reaction which does not merely react but acts anew, unconditioned by the act which provoked it and thereby freeing from its consequences both the one who forgives and the one who is forgiven."[20] This level of freedom is the freedom from the inevitable automatism of action and reaction, from stimulus and response—a cycle that would otherwise never come to an end. We experience true freedom when we are able to recognize that empathy equals destiny. When we can see the world through the perspectives of others, when we can imagine and relate to what they're feeling, when we can attribute mental states to them and truly feel they are as real as our own within the Mind of God, then we will truly be free.

Do Good and Evil Actually Exist?

Remoteness from God is not a matter of physical distance, but a spiritual problem of relationship.

—RABBI ADIN STEINSALTZ[1]

The individual is the human who is all mankind. The whole history of man is written in ourselves.

—KRISHNAMURTI[2]

It was Christmas Eve 1989, New York City, and the psych emergency room of Queens General Hospital felt like its own universe. As a second-year psychiatry resident (only just beginning to ponder a switch to neurology), I, along with my best friend at the time, a co-resident on the psych rotation, had already learned about this dynamic in the ER. The ward had a separate entrance and triple-bolted doors. One way in, and one way out. This was to protect volatile patients and vulnerable staff alike.

My friend was one of the funniest guys I ever knew, and he made bearable the rare dead nights like this one. Most nights in the ER were difficult, populated by sui-cidal teens, addicts on drugs they were not supposed to be on, and the mentally ill off the drugs they were supposed to be on. But this night was particularly slow. Near midnight, my friend sidled up to me with a grin and said, "All quiet on the Western Front."

I winced. Baseball players and psych residents have their own superstitions. Psych residents never said it was "quiet" in an ER—that was the kiss of death.

Sure enough, moments later, EMTs with two attendant police officers brought in a small, dark-haired, dark-skinned woman, no more than five feet tall. Despite her diminutive size, she was strapped to the gurney with four-point restraints (the modern and more humane equivalent of the straitjacket). All she could move was her head—which she reared up while she quite literally snarled and growled.

"She's possessed," said one of the nurses.

Possessed? I'd seen the horror movies, and although I didn't believe in the existence of a literal devil, I was inclined to agree with the nurse's assessment the more I observed this patient. She snarled and growled at us again. The sounds coming from her were not human. Yet surely there had to be a biological and psychological explanation for what we were witnessing.

Didn't there?

Dr. Jekyll and Mr. Hyde

The night grew even more bizarre.

The technique for properly restraining a potentially explosive patient such as this woman required that one arm be secured up above the head, and one arm down. Seconds after they wheeled the woman into the ER, she somehow freed that higher arm—this takes superhuman strength, mind you—and landed a roundhouse punch to a burly cop, nearly knocking him down.

"This one's all yours," the cop said, rubbing his contusion, and the two policemen scurried off, hitting the exit button to escape the ER.

The duty nurse buzzed them out while the EMTs got their paperwork signed, and I steeled myself for my first close encounter with the woman. As the resident on call, it fell to me to figure out the best course of action. The patient was so violent that right away I moved to administer 1 milliliter of Haldol via intramuscular injection. This drug tends to work wonders at reducing aggression and the desire to harm others, as well as clearing a patient's thoughts. But as I and two of our most seasoned nurses attempted to restrain the woman so we could administer the antipsychotic drug, she thrashed in her restraints and tried her hardest to bite my arm the way a rabid pit bull might.

Normally, at the same time that we take vitals in this intake room, I would also take a history. But this patient was still way too hostile, and with the exception of guttural grunts and the snapping of teeth upon our approach, she was still uncommunicative. It wasn't more than two minutes into the meeting, and already I was flummoxed about what exactly was going on. Fortunately, a triage nurse came through the curtain to tell me that the patient's husband had arrived and was waiting outside. We had administered the Haldol by then, so I took the opportunity to escape temporarily and meet with him while the antipsychotic was taking effect.

The husband paced in the waiting room and displayed signs of agitation. But this is not in itself unusual. In answer to my first few questions, he told me his wife's name was Magdalena. She was twenty-nine, a native of Guatemala, in good physical health, and that "this happens time to time with her." The onset of this latest episode had been five days earlier with no overt precipitating event. The husband told me he was angry with the police for calling an ambulance, and angry with the paramedics for taking her to the hospital.

"Why would you be angry?" I asked, aware that my tone revealed no small amount of incredulity.

"Because she doesn't need a doctor," he said in his clipped accent.

"What does she need then, sir?"

"She needs an exorcist!"

If a patient is distressed, I don't normally take this long interviewing a family member, but I knew that Haldol can take some time to calm a dangerous patient (not to mention that I was in no hurry to go back to her), so I sat down to gather more information, and I invited her husband to do the same, hoping to relieve his distress as well.

"Is she on any medications for schizophrenia?" I asked, working through a list of plausible conditions. "Bipolar disorder? Anything?"

"No," he said. "She's no crazy lady—I tell you."

"Did she hit her head at any point? A car accident? A fall?"

"No. No!" A vein in his temple pulsated with agitation.

"What about seizures? Do you remember if she ever—"

"Why aren't you listening to me?" He stood abruptly. "She doesn't belong here! She's not sick—she's got the evil spirit inside her! You take it out, or I take her somewhere else!"

I'm not ashamed to write for posterity that this conversation—especially this man's insistence upon the one and only sure diagnosis—shook me deeply. But I was also certain—not only because second-year residents tend to get cocky about their diagnoses but also because I genuinely wanted to help this clearly distressed woman—that admission to an emergency psychiatric facility was the obvious choice for her, as was my attempt to "medicalize" the condition. I was certain we could help her here, even though at the moment I wasn't sure exactly how. It seemed like I'd gone as far as I could with the husband, so when Magdalena was clearly sedated, I reentered her room. I have never been more fearful in a medical situation before or since.

And never more surprised!

What I encountered inside the room was the most meek and mild-mannered young woman that I could imagine. She displayed a countenance of complete kindness, humility, charm, and serenity. She even averted her eyes somewhat, the way an Old World servant might have done.

In a high-pitched, lilting burr of speech a little less reminiscent of her Central American Indian ancestry than her husband's, she said, "Doctor, I so hope I didn't cause any trouble to anyone."

That was the mother of all understatements. I was so aghast at the difference in personalities that I immediately locked eyes with both of the nurses in the room. They, in turn, both raised their eyebrows. Haldol is a powerful drug, but there's no way it can make such short work of transforming Hyde back into Jekyll.

The woman had complete amnesia of the events that had just occurred, but one of the nurses had already filled her in on why she was in the hospital. And that was the rub. I realized it hadn't been the drug at all that had done the trick. What I had on the gurney before me was a genuine—and severe—case of multiple personality disorder, or what's now referred to as dissociative identity disorder.

A brain dramatically transforming itself on a dime was one of the most fascinating processes I'd ever witnessed, albeit a terrifying one. And while still focusing on my patient, almost literally at that very moment, I decided I would change my career to neurology.

In the course of my perfunctory exam of the woman, I discovered that her heart rate, pulse, body temp, and other markers were dramatically different from the ones taken by the EMTs in the ambulance just a short time earlier. I admitted her to the locked psych ward on a standard hold with the diagnosis of Psychotic Disorder NOS (Not Otherwise Specified), as I needed to leave it up to more experienced professionals to confirm my multiple personality disorder hunch. In the psych ward she'd get a full workup.

Our chief attending psychiatrist, "Dr. S," a man in his

late sixties, took over her case. Over the next few days, I checked in with Dr. S from time to time about Magdalena's case and found out that her other personality, the one her husband called the evil spirit, had not manifested since she'd been admitted. But the vehemence of that personality's violence, the suddenness of its onset and departure, and the contrast between it and the woman's other, submissive self, made Magdalena's case very interesting to everyone at the hospital.

Dr. S told me he was determined to meet Magdalena's "other half." He had a plan for doing so—and all the psychiatry residents were invited.

In a Hypnosis Session

Dr. S's plan began with a routine consult to the standard workbook, the *Diagnostic and Statistical Manual of Mental Disorders,* published by the American Psychiatric Association.[3] The book locates the root cause of a number of serious mental conditions such as multiple personality disorder and dissociative identity disorder in serious, often sexual, childhood trauma.

One needn't search for a remote version of "evil." While there's some controversy about this diagnosis, many experts believe that childhood trauma can be felt so extremely in a person's psyche that it necessitates a fragmentation, a compartmentalization of the disturbances, so that the trauma is unavailable to consciousness. One dominant "part" of the person might remain aware and "normal," even while other "parts" have "split off" into different directions and remained traumatized.

A core part of the standard treatment strategy for multiple personality disorder and dissociative identity disorder is unifying the split parts of the self into one whole based on the dominant host personality.

Toward this aim, in collaboration with Magdalena's husband, and in addition to the psychotherapy she was receiving, Dr. S wished to include an adjunctive modality that often benefits patients like Magdalena: hypnosis. And Dr. S thought it would be instructive to conduct his first session in front of all the psych residents.

The session began with Dr. S gently calming Magdalena until she went into a trance. Her eyes rolled back. She answered all his questions in her polite and demure manner: "Yes, Doctor, I'm comfortable." "Yes, Doctor, I hear you just fine."

But when he asked her whether she was ready for him to address "anyone else" who might be there with her, she said, in her regular voice, but a bit more pleadingly, "Please, Doctor, no."

"No, what?" he asked.

She whispered: "We shouldn't let him out."

Whereupon, like a colt out of a paddock, "he" was out. The other personality was with us. It was as though that switch went off again, and the lights behind the woman's eyes went dark. Everything about the normal Magdalena fell away, died in a sense, and he, her other personality, "the Evil One," filled up her skin. A deep throaty roar came from her. Her body distorted. Her entire countenance looked in no way human.

The psychiatry residents immediately rushed to the far side of the room—some knocking over their chairs in the process—and piled into the corner. Dr. S remained in his chair across from Magdalena while the Evil One stared and slavered, its chin on its chest, and its nostrils flaring. Dr. S gripped the arms of his chair and kept (literally) on his toes in case he, too, needed to bolt for the door. He licked his dry lips and eked out the following demand: "Tell me what your name is."

No reply came in human speech. Only one long grunt. The sound of a caged bull before the stable master pulls the pin on the gate. Why had no one thought to restrain her first?

Whatever the cause or the nature of this aberration, I can say with certainty that never before or since in my life have I experienced the power of a diseased mind so profoundly. And that's saying a lot. What I've come to learn is that a person's mind can believe so strongly in the reality of a split personality that a physical body can actually manifest that reality within the same person. One personality might have high blood sugar, and the other, not. Handwriting, artistic talent, language fluency, allergies, drug reactions,[4] and even clinical visual acuity and eye shape and curvature might all differ from personality to personality.[5] One personality might require prescription glasses, while the other personality sees twenty-twenty.

Walking, talking, breathing, and heart beating—there exists a "biological self" that corresponds to the psychological self, thinking, believing, hoping, fearing. And as the latter shift under given circumstances, so do the former, accordingly.

We all understand this intuitively: Imagine your neurobiological state if you were in a comfortable room at home and had been sleeping peacefully for the past six hours. Now imagine your body state ten seconds after you suddenly awaken to the clatter of shattering glass downstairs and someone shouting, "Fire! Fire! Call the fire department! There's a fire in the building!"

Dissociation essentially follows this same logic of bio-chemistry following psychology, though it works at the extreme edge of the continuum. The body chases hard what the brain believes, so when the mind thinks it belongs to someone else, then "someone else," in fact, inhabits the body. This "other self," which can take up to a few minutes, or just a second or two, to manifest, will thus correspond to an altered heart rate and breathing rate, and other physiological markers that demonstrate first a clear disorganization, which is then followed by a reorganization, and from thence a new pattern typical of the new, emerging personality. One of the most salient and stunning transformations happens right on the face, via relative tension of the facial muscles and reshaping of features, to match his or her new identity within.

The entire unsettling event with Magdalena lasted ninety seconds, then Dr. S, clearly as horrified as we all were, brought Magdalena out of her trance with a snap. She stretched her neck muscles, wiped her mouth, and smiled at Dr. S graciously. Then she noticed all of us "white coats" heaped in the corner of the room. She looked down at her lap, embarrassed. "I asked you, Doctor, please don't let him out," she said. "I meant it."

Hardwired to Biology

The other doctors can't dispute what they experienced in that room. So how do we explain what happened to this woman? How do we break down this disorder on a biological and psychological and scientific basis?

The disorder of Magdalena's self-identity seemed to point to some deeper level of understanding about the origins of dissociation, perhaps toward a fuller understanding of human identity in all its divisions and collaborations as well. "Man is not truly one, but truly two," as Robert Louis Stevenson's Dr. Jekyll asserts—one part is "good" and the other part is the opposite of good. This he calls "evil."[6] I would like to suggest something else, namely separation.

Within Magdalena's case history, she indeed had experienced severe trauma as a child. In her particular situation, she had been raped and sodomized by a stranger. The trauma had created such a profound effect on Magdalena that her identity had effectively split in two. One part seemed to hold her true personality—the polite wife and mother. The other part held on to and even embodied the horror of her childhood trauma.

Dissociation is the brain's way of removing itself from a traumatic experience. Whenever that other personality—the traumatized personality—took hold of Magdalena, it was indeed as if he displaced her own true being. The external evil had wound its way so tightly into Magdalena's biology that the other personality effectively "took over" from time to time. Magdalena could not control this other personality, much the same way we can't control having severe diarrhea—it just happens no matter how hard we try to stop it. For Magdalena, it was easier for her true personality to hold on to a dissociative state than for her to experience the reality of reliving the trauma of being raped and sodomized. When this external evil took over, the true Magdalena left. She was gone. That little girl was no longer there—both during the initial episode of trauma and whenever she relived that episode.

Does evil exist? Absolutely. Magdalena's childhood trauma was the epitome of the definition of evil. For this adult who had attacked her, it didn't matter what absolute and utter horror his actions wrought upon her. His brazen disregard for another human being's uniqueness and godliness, replaced by a hollow, selfish, and savage energy is my definition of evil in the flesh. Not only was Magdalena's body violated; her soul, too, was torn asunder.

I am quite certain that there will be people who take issue with any attempt to explain the complete and utter transformation of Magdalena's identity in strictly biological or psychological terms, just as there are those who would take issue with any "spooky" metaphysical explanation for what clearly was the result of extreme sexual abuse. The same reluctance to do so would mean a similar resistance to explore whether any brain disorder has a deeper existential meaning beyond its biological underpinnings. Each perspective, whether biological or spiritual, has its own subjective validity. We need to explore this subjectivity in our efforts to understand human suffering, especially when we inquire about our emotional states and how our beliefs are both a cause of and consequence of them. We must do this. Biology is always about the study of life, and life is always about meaning. Just as in Magdalena's case, our biology and our phenomenology are not separate things. They really are connected, integral parts of human consciousness.

As a doctor and a scientist who recognizes the imperative of this subjectivity, I come from the perspective that evil is not something wholly outside ourselves. In Milton's *Paradise Lost,* Satan is unable to recognize the difference between the noumenal (the real, the ordinary, or the normative) and the phenomenal (the extraordinary), and he infects humanity with this same curse when he first deceives Eve, who'll bring about separation and division. The manifestation of good and evil does exist within our biology; we just call it by some other name. The evil that poor Magdalena was experiencing was organic. It discharged itself within her subjectivity, infecting her physiology. We could say the woman was crazy. We can call her disassociated. But the evil was real, not artificial or abstract.

Walls and Bridges

So how did Magdalena's story end? It wasn't entirely a happy ending. I was a resident back then, so I don't remember all the particulars. I know she was stabilized, put on medication, and sent home. I'm fairly certain her episodes continued for some time. Did we miss something as medical doctors? Perhaps.

At that time back during my residency, I was struck—it would be more honest to characterize myself as *devastated*—by the insubstantiality of the psychological foundation that underpins our personal identity, how our lives become unified or disjointed through the identification/dissociation of our relations, both within ourselves and beyond, meaning among one another. The disorder of Magdalena's self-identity seemed to point not only to some deeper level of understanding about the origins of dissociation, but perhaps also toward a fuller understanding of human identity in all its divisions and collaborations. We all have experienced trauma in our relationships, and the obvious defense is to build walls and not bridges to protect ourselves from further hurt. Returning to Dr. Jekyll's assertion that "man is not truly one, but truly two,"[7] does this mean the human soul is the front line for an allegorical "angel" and "fiend" locked in battle for mastery of the universe—and of our very minds?

For me, Magdalena's case stood as the most stark example of how the mind's supposed imaginary states become real, how the allegorical becomes flesh. For lack of better words to describe my experience, at that moment in the ER I caught a fleeting glimpse of the raw and brutal force of the mind's agency and its manifestation in human identity.

Certainly there are documented cases of split personality where a priest or minister has come to the hospital and prayed for a patient and the disorder has been effectively cured. The underlying trauma is repaired. As medical doctors, we don't prescribe prayer for patients, but maybe we should. Some studies have shown that "intercessory prayer to the Judeo-Christian God has a beneficial therapeutic effect in patients."[8] And this makes sense, as most diseases and disorders of the brain contain both biological and experiential elements. One explanation is that a person's belief system is so powerful that a trust in a God who would remove an evil from a person's consciousness could indeed be part of the path to well-being.

From a scientific/philosophic perspective, there is clearly a biological dynamic of unity and segregation, walls and bridges, a morality and an immorality within us. As Italian renaissance philosopher Giovanni Pico della Mirandola wrote,

> Firstly, there is unity in things, whereby each thing is at one with itself, consists of itself, and coheres with itself. Secondly, there is the unity whereby one creature is united with others and all parts of the world constitute one world. The third and most important (unity) is that whereby the whole universe is one with its Creator.[9]

We can see these dynamics in the most basic building blocks of life, our cellular makeup. Cells are actually incredibly complex structures that have incorporated ancient and originally free-living organisms in a process known as endosymbiosis. The *symbiosis* part of that portmanteau means that cells are actually mutually beneficial and reinforcing. They display a form of reciprocal altruism. Cells must cooperate and interrelate to evolve and form more complex organisms.

Cancer cells provide a perfect analogy of a biological evil, where the normal cooperation and integration of activity becomes disentangled, autonomous, and apart; and where all of its activity exclusively serves its own ends. Cancer cells are doing basically the same thing that the rapist who sodomized Magdalena did for his own gratification. These cells strangulate their host to feed themselves and themselves only. Cancer steals a person's blood supply, oxygen, and nutrients. It's like the cancer cells are declaring, "My essence is the only essence that's relevant." Its will to power is its salient feature. Its mission depends on quenching itself on vitality robbed from the rest of you. It has to dominate. It must vanquish the liver, for instance, to live. The nature of cancer is one of separation and division, of life setting itself violently apart. The paradox, of course, is when the cancer ultimately overthrows the host, it has to perish, alongside the host it has killed.

The Neuroscience of Karma

Within any discussion of good and evil, we must ask ourselves, Are we connected with each other in any deep and fundamental way, or are we all completely separate, autonomous beings? Those who hold to complete autonomy contend that we do not feel the reverberations of anyone else's actions. Those who hold to entanglement insist that our selves, like cells, are more joined than we may initially think. As such, we have a moral responsibility toward each other.

Our perceptiveness has evolved to a state where we have managed not so much to unite ourselves with each other as to objectify and commodify each other. This objectification is a grave danger, for it leans toward a transformation of ourselves and our world into objects of utility. We must continually battle this commodification. A simple benchmark for human morality must exist, and it must be based on the straightforward belief that humans are not objects or commodities. We are divine sparks, and there is an imperative Kantian morality that instructs us never to treat another human as a means, but always to see another human as an end in himself or herself. We cannot discover any truth whatsoever about ourselves without an understanding about how our actions deeply affect one another, even perhaps our own lifetime.

In clinical psychiatry and neuroscience research, study after study has shown that a vast array of external forces influence the brain and affect our behavior. Studies involving Holocaust survivors as well as U.S. children of war veterans have documented significant generational behavioral effects of trauma.[10] This passing along of trauma is how the brain's gene code, programmed for survival, becomes a self-generating instrument of further shackling, a "mark of Cain," if you will. The bottom line is that if a brain is altered, then that alteration can be passed along to subsequent generations. "We cannot live only for ourselves," wrote the nineteenth-century priest Henry Melvill. "A thousand fibers connect us with our fellow men; and along those fibers, as sympathetic threads, our actions run as causes, and they come back to us as effects."[11]

So we can extrapolate and thus imagine how we—each individual in this lifetime—comprise a part of a long, complex, causal chain. Not only our actions but also our words and even our thoughts affect others. Our lives are embedded in invisible but potent resonances, both biological and psychological, and those resonances move us and move others in turn. Resonance and harmony both begin and end with the aim of agreement, consistency, congruity, and the natural order of things. Good and evil are not simply actions propagated by individuals only. There is a case to be made for "all in one" thinking and movement: that our actions and behaviors affect swathes of humanity, perhaps even all of humanity.

Physics has shown that all of the universe is entangled. Particles of energy and matter can and do become entangled. What happens in one corner of the universe can affect something in another corner of the universe, many thousands of light-years away. Everything in the universe is in some literal contact at all times, and this simply escapes our wildest imagination.

Brains communicate with each other—our internal biological states are altered in both conscious and unconscious ways in a process called entrainment, patterns of brain wave activity coupled to our emotional and cognitive states. We see this entanglement through everyday social interactions, which can be both verbal and nonverbal. Consider when all the people in a social group laugh at a joke together. Or how people at a concert tap their feet in unison to the music. Sometimes, when talking with a friend, we can incidentally coordinate our speech frequencies and hand movements without even realizing it.[12]

In psychology, we use the term *resonance* in a similar way. It implies the tendency to enlarge and amplify through matching the resonance frequency pattern of another resonance system.[13] Examples of such resonance in psychology include mother-infant bonding, where a host of biological and psychological factors between parent and child become tightly coupled.

There's a reason for our synchronicity. Such connectivity is a biological imperative. We're not meant to be lone wolves. Our survival literally depends upon social bonds. We often experience this interconnectivity unconsciously, from the moment we as infants bond with our mothers. Or perhaps we've experienced this with our lovers. Like pendulums that swing in unison, this coupling of separate brains can connect us in moments, almost as though we're one. We might finish each other's sentences. We might take on the rhythm and cadence of each other's speech, and this synchrony has been shown to exist among good friends and family members. Our brains are programmed to develop this kind of evolutionary adaptation so we can maximize our success because we are social animals that respond to each other's behaviors.

There doesn't need to be anything spooky about these synchronistic links. These resonances also affect us on the biological level, including our heart rates and brainwave activity. Our brains continually produce waves that we can clinically measure as different types of electro-encephalograph (EEG) patterns. EEGs are resonance-detecting technologies, and we use these measurements of the electrical activity of the brain to diagnose epilepsy, for example. Every neuron in our brains resonates through fluctuating electromagnetic forces, measured by the EEG technology. We have hundreds of trillions of these neurons, and the electrical charges induce waves of unique activity, which help define our thoughts, memories, and behavior through these internally generated fluctuations of energy.

The early twentieth-century German neurologist Hans Berger was the first person to describe differing brain waves after reporting a "spontaneous telepathic experience" with his sister.[14] I consider *telepathic* to be a word that is too loaded to use today, so I prefer *synchronistic*—a type of unconscious entrainment where brain activity becomes coupled in individuals who share a deep emotional connection. For Berger, the synchronistic feelings occurred in tandem with a near-fatal cavalry accident he suffered while training with an artillery unit in 1892. During the incident, he barely escaped the oncoming wheels of a mounted cannon and nearly died.

When he returned to his quarters that evening, Berger found a telegraph from his sister, with whom he was quite close, inquiring about his health. That morning she'd been overwhelmed with a feeling that something terrible had happened to him. The timing of this telegraph surprised Berger. He hadn't heard from his family since joining the army some months earlier. He was convinced that the synchronicity of this event went far beyond mere coincidence—it "resonated" with a deep meaning for him in ways that would change not only his life but the history of neurology as well.[15]

Berger became obsessed with the concept that humans can communicate telepathically, especially at times of great stress or mortal danger. He devoted the remainder of his life to exploring the physiological bases of psychic energy, and the correlation between objective activity in the brain and subjective psychic phenomena.

What can we learn from these observations? We should assume that not only what we do but also who we are and what we think will reverberate through others and have an effect on their thoughts, feelings, and actions as well. This is, I believe, what is meant by *karma,* and we are now more apt to understand it by measuring the effects our actions have on others, not only behaviorally but physiologically, too.

An example of the power we have to affect each other in tangible, biological ways has been examined in many medical studies related to the effects of intercessory prayer, when a person or group of people pray for the emotional or physical well-being of another. While these studies, when examined in their totality, have been inconclusive, we have learned something very important about the effects of prayer on health. Many of the principal investigators of research studies involving the influence of prayer on those suffering a specific medical illness have concluded that intentionality can facilitate healing at a distance. While no strictly material explanation can account for the effects of prayer, the data seem to indicate our intrinsic connectivity and the agency of relationship. Whether we attribute the effects of these connections to God or to some natural force, prayer clearly exerts some measurable energy that is felt in the biological world.

Why Evil Exists

Few, if any of us, can sufficiently answer the question: Why would a benevolent God allow evil to exist? Some contend that it is virtually impossible to find anything benign about God when tragedy befalls us. Certainly it is difficult to find meaning in the countless ways we encounter suffering on a daily basis or to understand the significance of natural catastrophes or illnesses that afflict us or someone we care deeply about. We will never fully understand the reason for our suffering. Neuroscience can't help us answer the question of why we live in such a seemingly indifferent, precarious, and often brutal universe. William James, a great American philosopher who lost one of his children to illness, reflected on the unpredictability of our physical existence, which he described as "the pit of insecurity beneath the surface of life."[16]

The question of why evil exists as a continual plague in the mind of man is an entirely different existential question, something that is not external or exogenous to us in the same way as a natural disaster is. When confronting evil, it is not a question of whether God is absent from our lives. The nature of evil is human nature; therefore, the question of its existence is for ourselves, not God, to answer. Evil is, at the end of the day, an issue of how and why we ourselves are removed from our own humanity. Our answer regarding the existence of evil is found in our response to the question, not to the question itself.

We can only respond and confront evil with compassion filled action, not contemplation. Thus we must respond with whatever capacity of goodness we have within us. We are certainly burdened with the responsibility of faith: to offer evidence to ourselves, through the activity of our own lives, not only that God exists, but that his true nature is benign. The only way we can reveal this is through our relationships with each other, to be placebos of sorts through love and unconditional kindness. The word *religion* comes from a Latin root that means "to bind together."[17] At its best the truly religious experience connects us to others. A false religious experience divides us.

When Moses asked God to reveal his name, God simply said, "I AM that I AM."[18] There is a Hebrew translation that is interpreted to mean "As you are with me, so I with you."[19] It is in our in-betweenness where God is found. Morality is not beyond us; it is fully within us. Our faith is found in our deeds, our love, and our caring for each other. That is where we find God's response to our pain and our response to the perceived divisions between us.

Immortality: The Remembrance of What Is

We shall not cease from exploration
And the end of all our exploring
Will be to arrive where we started
And know the place for the first time.

—T. S. ELIOT[1]

One of the biggest questions all of us human beings will need to ask—and answer—sometime in our lifetime is this: What comes next? What happens after we die? Which then begs the question: Is there something beyond this mortal coil? Do we experience some form of afterlife, or does the end of our earthly existence mean lights out for eternity? This question begs others: Is there a purpose in life? How is our search for meaning coupled to the question of God's existence?

Most faiths, if not all, define the experience of an afterlife as an encounter with God. Alternatively, without the possibility of an afterlife, we confront the brutal alternative, as Milan Kundera writes in *The Unbearable Lightness of Being*: "a life which disappears once and for all, ... like a shadow, without weight, dead in advance, and whether it was horrible, beautiful, or sublime, its horror, sublimity, and beauty mean nothing."[2]

We can approach the question of immortality in any number of ways. Some arrive at agnosticism—that the only certainty we have is uncertainty, and the only way we will know what happens after death is when we take the journey ourselves. Others take the view of a nihilist or an annihilationist—that there is no life after death, or that all life is simply snuffed out, and all that awaits us on the other side is a big blank void of nothingness. Others hold to some sort of literal heaven and hell—that either an existence of consciousness and bliss awaits, or an existence of torment and/or separation from God—and that these states continue for eternity. Still others hold to a variety of hybrid views. In Judaism, for example, the concept of an afterlife is regarded as "the world of understanding," where the illusory realities of our physical existence are unveiled.

How would we answer the question of immortality from a neuroscientific view? The entire endeavor of arguing for an afterlife or any concept of immortality hinges on a neurobiological argument about the essence of our "selves." In this exploration we are referring to the mind or soul. The question is: Where are "we" within—and indeed, without—our bodies? That discussion begins with a recognition that the brain and body are fundamentally interdependent. So, too, are our technologies, cultures, and religions—all are extensions of our brain's inner workings projected outwardly.

But, as I've been arguing, we must also consider the mind, which is not strictly the same as the brain. Our souls/minds are in essence immaterial, extra-somatic if you will, that is, beyond the functioning of the body. Yet is there any objective data to support what happens when the mind is unhinged from its biological constraints?

Can we ever know for certain what happens to us when we die?

The Significance and Anatomy of Memory

Perhaps our afterlife is solely one of memory. All memories fade, and some memories seem to disappear altogether, but there is always a way to resurrect a memory, to bring an event "back to life," in a sense. So memory becomes key to our understanding about the nature of immortality. As Ewald Hering, a Viennese physiologist who was known for his studies on vision and perception, said,

> Memory collects the countless conscious phenomena of our existence into a single whole; and as our bodies would be scattered into the dust of their component atoms if they were not held together by the attraction of matter, so our consciousness would be broken up into as many fragments as we had lived seconds but for the binding and unifying force of memory.[3]

Memory, in all its various forms—biological, cultural, spiritual, and psychological—is what holds things together. Without memory, it is like our histories have no meaning, no enduring legacy. Maybe this is what the poet W. B. Yeats was referring to in his prescient 1919 poem "The Second Coming": "Things fall apart; the centre

cannot hold; / Mere anarchy is loosed upon the world."[4]
Could it be that different forms of amnesia lead to things
falling apart? Are we going through a period of history
in which we have forgotten who we are? Are all this war,
violence, turmoil, racism, sexism, and fear symptoms of
a collective amnesia? And could we all find redemption
when we start remembering again?

Yet when it comes to believing in an afterlife, there
are more than poetic metaphors at stake. We must place
this idea in the context of what we know about the laws
of the conservation of energy, which states that matter
and energy can change forms but can never be destroyed
or created. A human life itself may disappear, but what
happens to the energy within that life? What happens
to the consciousness of a person's mind? The energy is
not destroyed. As we know, energy can't be destroyed. So
the energy that makes up a life must still exist "somewhere" after a person dies. And perhaps this somewhere
lies in a wider understanding of memory, not necessarily in a strictly physical form but in some other, inter-
changeable form of energy. We humans are much more
than mere shadows; and there is much more to life than
nothingness. Undoubtedly our energy is simply scattered
after death, like heat let out of a room when the door of
a house is opened on a cold day. Yet perhaps that energy
can be "remembered" in our collective minds.

From a sheer biological perspective, we know that a
person's memory has a topography of sorts. Compressed
within the brain's three-dimensional structure are more
than a hundred billion neurons, a million billion synapses, and a hundred trillion connections, a vast forest

of trees in the cave of our cranium with more junctures than there are stars in the universe. Our neurons, "the crown of evolution," as they have been called, infuse our brain like microscopic threads through our cortex—a wrinkly, wet sheet folded in on itself again and again.

The synapse, the fundamental site of biological information processing (think of it as the original information superhighway), is the extremely small gap where the nerve impulses that bear all the information of our biological existence are transmitted and received, via axons and dendrites. These networks create an intricate web of connectivity that governs our thoughts and memories, and in the subtle contact at the hundreds of trillions of signaling junctions of those synapses, where electrical charges and minute waves of finely balanced ions operate, all our thoughts, memories, personalities, and our dreams are born and live in the domain of created time.

Brain cells in the hippocampus (a spiral, seahorse-shaped fold of gray matter) act as the principal repository and storage house of memory, where anatomy equals destiny. Here resides within us the archives of our personal history, the area of the brain that records and embodies the flow of time. Think of the hippocampus as our cognitive map, guiding and connecting our social memories—the emotional and experiential territory between us. Within this pyramid (described as such because of the unique shape of the hippocampus), the neurologist is like a cartographer looking to chart memorable destinations in these cognitive maps. Our memories create lasting traces, chemical footprints of experience, called *engrams*—permanent impressions within brain cells.[5] A specific memory—your first kiss as an example—is en-coded through a process known as synaptic tagging. Spe-cific proteins become "charged" with experience. It is the transformation of these tagged proteins that allows us to process time within our brains.

Every human life has its own unique history, just as every culture has its own unique history and memories attached to it. This is why we in America celebrate the Fourth of July. It's a way of commemorating a fixed point in our national memory. Yet to grasp this history, we must also have a sense of time, like the hours of a clock, the sequential passing of events, like all the events that have brought us to where we are today as a country. We may only have a vague memory of some of these events or no memory of them at all, but still past events have forged where we are today, for better or for worse, as a people. There is no life without time, and no time with-out life. According to the physicist Julian Barbour, "The brain is a time capsule. History resides in its structure."[6]

Thus, memory, time, and history are all closely inter-twined. Memory can only exist within time. Time gives structure to memory, much the same way our body gives structure to our thoughts and to our soul. Time is past, present, or future—something happened, is happening, or will happen, and events always occur in some chronologi-cal fashion. Yet the notion that time is a process that flows directionally may be an illusion designed within our own brains. For example, many people in life-or-death situa-tions perceive an altered rate of time. Significantly, events either seem to speed up or slow down. Traffic accident victims sometimes talk about how "everything seemed to happen in slow motion." That's because time is actually a perceptual property that does not have any objective existence. There is no time outside of consciousness.

If that weren't enough to chew on, we need to factor one more function of the brain into the equation—dreaming. The Talmud says that sleep is akin to one-sixtieth of death.[7] What does this mysterious statement mean, and does it provide us with a clue to the possibility of an existence beyond our physical bodies?

Memory, Time, and Dreams

Dreams are timeless and yet essential to our memory. In the dream state, for instance, where the dynamics of sleep are generated by the lower brain stem, we find no formal sense of time. Our minds seem to abandon any need to discriminate the "real" from the "impossible." When we sleep, there is no predictable cause and effect. The experience of dream states is most distinguished from the wakeful state in that we have no ordinary sense of time in our dreams—there are no beginnings or endings—and the terms *now, then, before, after, during, earlier, later,* and the like are not relevant to the unfolding of narrative or the experience of the dreamer.

Freud insisted that dreams represented the "royal highway" to the unconscious.[8] He intuited, without the benefit of our current tools to study our brains, that the flow of theatrical plots was unhinged from the normal wakeful dynamics, and that this unhinging originated from the depths of the soul. These sleep-oriented neurons, freed from the constraints of the hippocampus, were able to generate an experience for our brains that seems timeless and closer to our true, uninhibited essence.

Neuroscience's current perspective is that dreams represent a secondary phenomenon, or epiphenomenon. They are like exhaust fumes from the engine of the human brain, a random discharge of psychic energy without any intrinsic value or meaning. But I don't completely buy this viewpoint. It's too reductionist, and it minimizes the meaning and value of our dreams. If we only see our dreams as engine exhaust by-products of our daily lives, a sideshow to the only meaningful action, which occurs in our wakeful states, that viewpoint erodes our trust in, and our valuing of, our own interior and subjective states. William James cautioned against the dismissal of subjectivity from science, writing that individual experience "is infinitely less hollow and abstract than a Science which prides itself . . . on taking no account of anything private at all."[9]

While consciousness is seemingly destroyed in death, it is similarly obscured by sleep. The connection of our dream states to our spiritual essence, that which is beyond time and independent of the constraints of physical reality, is why the Psalmist writes, "When the Lord will return us from our exile, we will have been like dreamers."[10] This can be taken to represent a mysterious transformation from the cognitionless primal state awakening into life and consciousness.

What's the message for us? That there is life beyond ourselves, a life we can't measure, quantify, or ever grasp with our mostly limited minds. We get a glimpse of an alternative reality, a brief peek beyond the firewall that obscures our higher vision, when our civilized brains go offline. We know of another reality, beyond ourselves, and it's a reality saturated with meaning and truth that are entirely different from the ones we have grown accustomed to. Dreams, time, memory, and immortality are all connected. But how are they connected? What is the invisible glue that binds our existence, our dreams, our memories, and our lives?

Within perennial philosophy, our emotions play an integral role in how we experience and interpret the meaning of our lives. Our connections to ourselves, to the universe, are stitched together by the weaving of hued emotional threads through the taut fibers of black-and-white thought. We construct the meaning of experience—its *signification*—based upon the assignment of *values,* all of which rides upon the vehicle of *emotion.*[11] In the wide chasm between mechanistic biology and the immaterial soul, it is our affective states, and not our cognitive ones, that bear the true gravity of our existence. Our feeling states are much closer to our core and essence, that which is nonquantifiable, nondeterministic, irreducible, and ineffable—more fundamental to our sense of self than all our other "body parts" combined. Note what Jean-Paul Sartre wrote on the "magic" of the emotional processes in the construction of what we're wont to call reality: "The emotion signifies, in its own way, the whole of consciousness. . . . It expresses from a definite point of view the human synthetic totality in its entirety."[12]

Emotional and feeling states are an invisible metaphorical glue that binds us. Like motion in physics, our feeling states are produced by the action of this immaterial force, the core signification of our being, upon objects, that is, our flesh and blood. Einstein's elegant equation of $E = mc^2$ signified the interchangeability of the fundamental facets that make up our reality, proving that energy and matter are, as eighteenth-century British poet William Wordsworth would say, "deeply interfused."[13] All of this is to say that we are much more than our physical bodies and what we can measure. It is the heart at the center of the universe, our emotions, which bonds us not only to ourselves but also to the universe at large, not only in this life but also in the life of the world to come.

Life Beyond the Grave?

Death means the physical loss of a being. When a person dies, his or her biology ceases. But just because the biology ceases, that doesn't mean that the dreams, the emotion, and the consciousness cease also. As I mentioned earlier in this book, my father had a stroke when I was nineteen. He was a beautiful man, funny, intelligent, and devoted to his family. He was worried especially about me. I was directionless in my teenage years—experimenting with drugs, hitchhiking to Grateful Dead concerts—and I dropped out of college to find myself. My father's stroke left him profoundly disabled. Gone was the man I knew who told jokes laced with irony and self-deprecation. Gone was the foundation of my security.

After his stroke, he was transferred to a veterans hospital, where I would visit him. These encounters were painful, as he no longer recognized me. His roommate was a mentally impaired veteran who would endlessly ask, "Does anyone know what time it is?" And over and over again my father would have to listen to this. No matter what the answer was, the question would be repeated ten minutes later as if it had never been asked. Whatever humor was left in my dad after his stroke was one afternoon conjured up with his response: "Will someone please give that man a watch, goddammit."

It was almost exactly a year after his stroke that my father died. After my father's death, I would dream of him on a regular basis. The plot of my dreams almost always unfolded the same way. I discovered that my father had been abandoned somewhere, a dilapidated warehouse perhaps, alive but bedridden and totally disabled. In the dream, he had been neglected or forsaken for so long that most of his faculties had vanished. He didn't recognize me or even acknowledge that I was his son, even though I pleaded with him and tried to wake him from his stupor. He just stared up at me with eyes that evinced no spark of recognition. Words couldn't describe the desperate agony these dreams elicited as I dreamt them, or the residual pangs I continued to suffer, sometimes for days after I had them, as though my father really were trapped somewhere, unable to communicate, and I was his only hope of salvation. Liberation was his, if I could only forge some key to release him.

Many years later, I had the first of what became a different kind of recurrent dream about my father, with a quite divergent tone from the ones I'd endured decades earlier. The dreams began the same way, though. With the new dream, again I found my poor father in a dilapidated warehouse, alone, emaciated, and bedridden. I said, "I'm here to take you home, Dad." But this time, there was more than a mere ember of consciousness behind his eyes. There was love for me, his son. He looked at me with an expression of utter relief, with gratitude and kindness, and he said, "I've been waiting for you for so long. Oh, I'm so happy now, Jay!" I lifted him up from the decrepit bed, and cried, releasing years of pent-up fear and anguish as I carried him across my shoulders to freedom. I could feel his presence vividly—his sharp bones, his weakened muscles, the sweet, sad smell of his ramshackle body. And as much as I wept, he laughed, as though he were lighter than air.

To this day, I believe a shift in my consciousness led me to a new awareness of my father and my relationship with him.

The unconscious represented to Freud a repository of repressed memories that can profoundly influence our thoughts, beliefs, and actions.[14] Surely dreams are never actually forgotten. They only disappear because we are unable to recollect them. Energy is never lost as we've mentioned, so our dreams are buried somewhere deep in our brains and deep in our minds. Those dreams are still accessible, still real, still productive, still molding our lives in ways we are not fully conscious of. Isn't that evidence that points to something beyond this mortal coil?

The day after I first had this liberating dream, I was driving to a medical conference where I was scheduled to speak. But unexpected roadwork detoured me off the highway onto a smaller state road. I followed the orange detour signs and the cars in front of me awhile, but as I was reviewing in my head the wording of my lecture, I suddenly realized I was lost. I made a few turns in the direction I sensed the highway was, and then I passed a cemetery. The name on the sign over the cemetery's entrance rang a vague bell. I called my sister, and she confirmed that this was the cemetery where my father was buried. I had not been there in many years, and I never would have come upon the place had it not been for the serendipitous coincidence of the road detour.

I pulled into the gravel drive, parked, and spent a long while trying to find his small plot. When I at last did, I sat there staring at his diminutive gravestone and the surrounding few dandelions, the only proximal, physical memorial of his life. The gravestone read, LOVING HUSBAND, FATHER, AND GRANDFATHER. That was it, the entire contents of my father's life—all his hopes and dreams; his marriage, children, and grandchildren; his work, rest, and play; his shortcomings and contributions—summarized into a few short words that my siblings and I had chosen many years ago.

When I saw them again, those words at first seemed puny, and I felt guilty for their glibness, their insubstantiality. Yet when I read them back and contemplated their value, I realized these holographic words etched in stone said all that needed to be said about my father. And my father's voice could be heard. Not in audible words, but in a sense I received in my heart: "I've been waiting for you for so long," he said. "Oh, I'm so happy now, Jay!"

That was an ineffable experience but one that many of us have experienced in our own ways, and while it may seem like a leap of faith, it was one that convinced me that there is an afterlife. And in considering those moments in the cemetery I find myself coming back to the premise of this chapter—and indeed, the premise of this entire book: "What is essential is invisible to the human eye." Death means the physical loss of a human, but it doesn't mean the being that is part of that "human being" dies. My experience in the graveyard awakened me to this truth. My father had not stopped existing. Sure, his body had died, but the totality of himself—that was not gone.

Of that, I, a man of science and a man who struggles with faith, am convinced.

A Life Without a Memory

Memory is often not very trustworthy. It is not immune to illness, injury, fraudulent transactions, and even unauthorized deletions. For instance, we tend to romanticize the past, and we often develop selective amnesia, picking and choosing the narratives we use to describe the events in our pasts. We not only do this personally; we do this culturally, such as when we glorify the outcomes of our wars. This is normal, and to be expected, yet memory takes an altogether different tack when coupled with brain disease. Nowhere is the vulnerability of our memories more fully on display than in patients suffering from Alzheimer's. We can learn a great deal about our immortal states when we see the overt loss of memory in a person.

I gave a talk in Aspen once on the biology of Alzheimer's, and afterward I met a gentleman and student of Buddhism who asked if I would treat his teacher, an elderly monk who was originally from Tibet. The monk would go on to become one of the patients I was most fond of and humbled by. After a series of phone calls from his polite young caregiver, the venerable monk was at last led into my office. The monk went by a single name—Tenzin—and his caregiver informed me his English was quite good.

"It's an honor to meet you, sir," I said to the monk.

"My honor," he replied, bowing at the same time he took my hand with both of his. I had to grin quietly. I'd had grateful patients before, but none had ever bowed to me, especially before I'd even treated them.

I quickly learned that Tenzin was an affable seventy-year-old, despite his progressive dementia, which his caregiver-nun, Ani Pema, told me everyone assumed was Alzheimer's disease. Those closest to Tenzin had first noticed the disease coming on about two years before. Everyone was now very concerned about his health. The disease seemed to be advancing, and his forgetfulness was becoming more obvious. They were in desperate need of an evaluation and recommendations.

As Ani Pema explained all this, I noticed that Tenzin continued to bow to my staff members as they came in and out, or even as they passed by the door. He smiled at everyone, and he looked in wonder and astonishment at the everyday items on my shelves and desk. I was aware that I was stereotyping and falling into cliché, but he seemed a man of genuine joy and pure spirit, unencumbered by the trappings of pettiness and materialism. He was particularly taken by the photograph on my desk of my two children, so I handed it to him as I listened to Ani Pema. Tenzin's eyes twinkled as he gazed at the children, and with his head cocked, he seemed on the brink of some delightful remembrance. Then a frown overtook his face like a cloud over water, and I could tell the memory was gone.

Sure enough, it was clear to me soon into his examination that he suffered significantly diminished short-term memory, a sign of an Alzheimer's diagnosis. I scheduled a series of subsequent evaluations and was able to confirm this diagnosis with some certainty. During that time I became much closer to Tenzin and even looked forward to his visits, which I confess I wouldn't say about all the patients I've seen. I always try to stay as objective as possible with my patients—an overemotional, overinvolved doctor is no good to anyone. But even Ani Pema could tell that delivering the diagnosis to this special man was a very difficult thing for me to do. Tenzin was surely venerated and had touched thousands through his dharma teachings. But he hadn't said a word to me about Buddhism. He just "was," in true Zen fashion. He was content with his existence. That model of a gentle, intelligent being somehow made me a better person, just for being around him.

It is always difficult to communicate the findings of an Alzheimer's diagnosis to a patient and his or her family. Loss of the ability to remember is one of the most dreaded afflictions we can suffer. Many call Alzheimer's disease "cruel," perhaps rightly so. As many of us think and feel, our identity and sense of "self" completely depend upon these recollections of the events of our life, our history and how it defines us. Alzheimer's disease robs its victims of their individuality. Without our relatively stable memories, we must grapple with the question of identity: Who are we and where do we stand? All the bliss and trials that constitute our lives, our friendships, loves, and losses, what we once experienced as the steady, forward flow of our lives, gets swept away like so many fragments of shells dragged and dropped by the merciless undertow of this disease. And what's worse, for a few of those tides, we are aware that the disease is befalling us, as it surges on us gradually before the unrelenting sweep becomes a tsunami. Like children whose hands are too small, too slippery to grip and the current too quick, our memories can be felt as they slip away into the deeper ocean. We can hear them for a while. We can see their arms waving. Then they're gone. And when our memories are gone, so are we. Standing on the shore of oblivion, all alone, strangers even to ourselves.

Memory is ego, and without ego there is no I. So in the absence of memory, ego vanishes as well. The unique tragedy of Alzheimer's disease is that it robs its victims of their singular nature, erasing the hard drive of memory that stores "self-memorial-hood" across the brain. The obvious paradox, though, in a case such as that of the Buddhist Tenzin, where there is a belief that ego is an illusion to transcend anyway, is this question: What purpose is there in resisting the dissolution of one's memories of pain or pleasure? The ultimate goal for a Buddhist is Nirvana, the deliverance from the mind—no longings, no wishes, no strivings of intellectual life. This represents utmost purification and enlightenment. As the Buddha taught his students:

> When ignorance is abandoned and true knowl-
> edge has arisen ... then with the fading away of
> ignorance and the arising of true knowledge he
> no longer clings to sensual pleasures, no longer
> clings to views ... no longer clings to a doctrine
> of self. When he does not cling, he is not agi-
> tated. When he is not agitated, he personally at-
> tains Nibbana [Nirvana].[15]

Tenzin seemed to understand this paradox, and even embrace it. He seemed perfectly comfortable that his brain was hitching a ride on a disease process that would insist—with the full force of biology—that these illusions would vanish without his effort.

On the third visit, I did another battery of tests. By this point, Tenzin had begun to conflate and scramble the time periods and people in his life, as Alzheimer's patients often do. Remarkably, though, these relationships were essentially anchored not in memories of resentment, but in compassion and fondness. After a discussion with Ani Pema, I prepared to explain to Tenzin in detail his diagnosis and its implications. He sat across my desk, beaming as usual. I did my best to explain in layman's terms what was happening in his brain. His eyes evinced that said brain was clearly absorbed in our dialogue, and his appropriate nodding and eyebrow raising told me he did not misconstrue in a gross way any critical point. When I was done with my elucidation of the Alzheimer's process, I asked the elderly man what he'd understood, and what he thought this diagnosis would mean for him, his vocation, his calling, and his future—and his answer surprised me.

"This is okay; it is fine. No one should be concerned with this matter," he said.

I turned to Ani Pema to make sure I'd heard right. She whispered something to Tenzin; he whispered back in a language that wasn't English, and they both turned to face me. Ani Pema said, "He says that he is grateful for your assistance, Dr. Lombard, but that it is not necessary to feel that there is anything necessary to do."

"But why?" I answered.

The monk spread his arms out wide, the robed arm and the unrobed one, so that it appeared he had just one yellow wing. "Now—everything," he said. "Beautiful—and new. All I see. Yes, all very lovely."

He looked happy almost, definitely contented, like he must have looked when he was a small boy, secure and at rest with his mother back in his village. I had never heard a reaction to an Alzheimer's disease diagnosis as paradoxical, as humble, as acquiescent as this.

After work that day, Tenzin, I, and a friend of mine, a professor of philosophy, took a walk by a lake not too far from my office. It was a cold winter's day, and the lake had frozen over. I was interested in learning more about Buddhism, especially from such a venerated teacher.

"No, I am not concerned about losing my memory," he told us. "This is an opportunity and not a deficiency. It is allowing me to experience all this without cover. Memory is like these clouds that cover the lake, so we cannot see its true reflection. I now see everything, and everything is quite beautiful here."

Unfortunately, the day at the lake proved one of the last few days of peace in Tenzin's life. With each of Tenzin's follow-up visits, this biologically assisted form of enlightenment began to feel more like a nightmare instead, obscured by the day-to-day behavioral realities the disease had wrought. Tenzin became "difficult," according to Ani Pema. The report one month: He wandered at night. The next: He refused to dress or eat. Soon he denied even his essential medication. Finally, he even opposed his loyal caregivers. On the follow-up visits, we needed to address these issues directly.

"Of no matter," he said in a cogent, rather Buddha-like defense of his stance.

"But shouldn't you consider the decision to give up caring for yourself in the context of your loved ones?" I asked. "Aren't they all dismayed by your dwindling desire to live? Aren't they all connected to you in our mutual continuum of life?"

He nodded solemnly at this.

My ego smiled and I added, "So you'll agree to eat and take your medicines?"

He nodded again.

"There's one more thing," Ani Pema said. Sheepishly, with her head down, she whispered to me, "He won't wipe himself."

"Are you forgetting to clean yourself on the toilet?" I asked him.

"No," he said, motioning for Ani Pema's ear. She

shrank back in obvious embarrassment. With a flick of his fingers he signaled to her that she should translate.

"He says he sees the Buddha . . . everywhere."

"I see," I said, and took this as an opportunity to further test my chops in Zen paradoxes: "Venerable sir, if Buddha is everywhere, even on your toilet paper, then even performing this simple obligation every day can be regarded as an act of holiness."

Tenzin smiled, appreciating the koan-like manner of our repartee. He seemed to grasp that his personal encounter with Alzheimer's disease was not the proverbial "sound of one hand clapping"—it was influencing those around him, too. He left the office that day agreeing to obey his caregivers' requests, and I felt satisfied that I had scored a small victory.

Alas, he forgot this pact almost right away, and his problematic behaviors became even more intense. Tenzin, as everyone knew him and as he had known himself for seven decades, was drifting away. No longer could anyone around him manage with reason and patience his emotions and consequent behaviors. We needed to address them pharmacologically. But the drugs employed to manage Alzheimer's are limited, and over the next few months, Tenzin's memory gradually disintegrated even further. I got the impression that some pitiless force had taken the book of his life, torn out the pages, crumpled them, and set them aflame. With that degeneration of the narrative of self, his manner devolved into a nearly uncontrollable state. This once docile holy man regressed into a sometimes hostile and angry state. In short, he was no longer Tenzin.

The Integrity of Our Lives

Alzheimer's disease is an ungluing of our implicit bonds, a dismembering rather than a remembering. In this disease, perhaps the irretrievable loss of memory and organic disconnection to our past is not only a physical disease of the brain but an existential one of the soul. I wholeheartedly agree with Pope Francis that we are suffering from a spiritual form of Alzheimer's disease.[16] Indeed, there seems to be an unrecognized willingness to purge the memorial record of our dark history, our misdeeds and misdirections, to organically bury our worst propensities. Is it so unreasonable to ponder a mechanism of social degeneration of memory, spurned by our exposure to so much wanton and ceaseless violence that our only two choices—complete desensitization (numbness, dissociation, irresponsibility) or devolution into wholesale warfare within or between our selves?

The integrity of our inner lives is surely becoming eroded by technologies that are leading more of us to live our lives in a virtual reality, while our true minds and hearts wilt and starve. Too many people have now become almost exclusively reliant on relationships based upon a digital universe where hordes of "selves" experience their lives in virtual isolation—separate and radically disconnected—a world where *friend* has become a verb with little or no meaning. Even as the Internet has offered opportunities to bring us together, we can argue that this new age has also unbound and disengaged us from real and meaningful interactions. We spin out our gossamer threads, but do they stick to anything or anyone? The irony of living at a time of meta-entanglement and connectivity is that much of what has been wrought is radical isolation. This is the splitting and devaluation of the godly sparks that exist in everything, the rejection of the significance of other lives created by God with purpose.

How do we find our way home again? How is the preservation or even the resurrection of memory connected to our redemption? Can neuroscience offer us any insights about the connection of memory to our immaterial existence and the possibility of an afterlife?

Ideas persevere in our brain and disperse through other minds as memes. Memes are the pollen of human thought; they replicate in our brains although we are unable to quantify them as such. We see evidence of the reproductive capacity of memes every time we find ourselves repeating key phrases like "at the end of the day." The memes of memory are a form of neurological propaganda—the legacy of sentinel experience that, once rooted, can kindle and spread through technologically based vectors such as social media. Great figures of history are immortalized in our brains as memes—these lives are, as Don DeLillo has written, "dead stars that still shine because their light is trapped in time."[17]

The memory of our shared origins, the evolution of our minds, and the freedom to determine our fate can endow us with trust. What is it we can choose to remember particularly? "As if reason were the only way we could learn!" Pascal proclaimed.[18] Which memes are so urgent that our lives are dependent upon our ability to reincarnate them? Is it simply the experience and memory of love that can provide us with the security we yearn for in a world that is otherwise ephemeral, brutal, and contingent? Like the ubiquitous forces of gravity, love both aggregates and coheres us; it instills order and meaning within our minds and brains. It may seem to us an empty space, invisible and immeasurable, impenetrable to our physical senses, but fortunately we have the capacity to restore a dimmed light inside our sentient but scattered fragments, and together, this restoration can spark the light of the fundamental Whole. It is an immortal light that doesn't fade. It resides outside of time.

An ancient Jewish mystical insight figures prominently as an account of creation. The chronicle unfolds like this: In order to make room for the universe of his infinite creation in a finite physical universe, God had to "contract" his Divine Self, concealing his essence.[19] So upon this celestial retrenchment, Divine Light needed to be contained and hidden for a time. There wasn't room yet in the universe for it. Therefore, God stored it temporarily in special vessels, which eventually shattered, unable to hold the divine energy, thus scattering it through the new universe.

The hidden reality is that we are the divine sparks of "not merely a piece of the entire existence, but in a certain sense the whole."[20] Our unique creation is God's way of choosing the means of expression of his will, our existence. Through our individual and collective human actions, as souls in bodies, we have the capacity and responsibility to restore these dispersed fragments of light—this love—back into our existence. It is thus only this, the realization through action that our essence is immortal. That essential element of who we are, that which is part of God.

This is how memory leads to redemption:[21] We exist because of God, and he exists in tangible ways through us. If we can envision the Mind of God—that our existence is contained within the thought of a transcendent being—then we can have assurance that we will continue to live through him.

This leads to a tremendous possibility. In the words of nearly each and every revered tradition, we must—and there is no room for compromise—love our neighbor as ourselves. We must see that our neighbor is one with us and one with God. We have been put here on earth for a reason. Our purpose is born within our biology. Our understandings of God are not meant to promote infighting or separate us.

Instead—and this bears infusion deep within all our collective memories—God knits us human beings together in love. It is a memory that goes back to the creation of the universe.

And it is in love that we live forever.

Notes

Chapter 1: The Mind of God

1. Stephen Hawking, *A Brief History of Time: From the Big Bang to Black Holes* (New York: Bantam Books, 1988), 175.

    "Mind of God" is a difficult title to start off with for any work, and I am humbled to offer my thoughts on the subject.

2. Ipsos Global @dvisory, "Supreme Being(s), the Afterlife and Evolution," http://www.ipsos-na.com/news-polls/.

3. Antoine de Saint-Exupéry, *The Little Prince,* trans. Katherine Woods (1943; repr. Waterville, ME: Thorndike Press, 2005).

4. David Chalmers, *The Conscious Mind: In Search of a Fundamental Theory* (Oxford: Oxford University Press, 1996), xiii. Chalmers coined the term the "hard problem of consciousness" to describe the gap between materialism and consciousness. He writes that there is no nonphysical evidence for the existence of a soul.

5. "Thomas Nagel: Thoughts Are Real," *New Yorker,* July 16, 2013; Thomas Nagel, *Mind and Cosmos: Why the Materialist Neo Darwinian Conception of Nature Is Almost Certainly False* (Oxford: Oxford University Press, 2012).

6. Max Planck, Albert Einstein, and James Murphy, "Epilogue: A Socratic Dialogue," in *Where Is Science Going? The*

Universe in the Light of Modern Physics, trans. James Murphy (New York: W. W. Norton, 1932).

7. John Polkinghorne, *The Polkinghorne Reader: Science, Faith and the Search for Meaning,* ed. Thomas Jay Oord (London: SPCK and Templeton Foundation Press, 2010).

8. Ibid.

9. Immanuel Kant, *Critique of Pure Reason,* trans. Norman Kemp Smith (1781; repr. New York: St. Martin's Press, 1965), 521.

10. Thomas Moore, *Care of the Soul: A Guide for Cultivating Depth and Sacredness in Everyday Life* (New York: HarperCollins, 1994), xix.

11. Jonathan Sacks, *The Great Partnership: Science, Religion, and the Search for Meaning* (New York: Schocken Books, 2011).

    Rabbi Sacks's book is a thoughtful exploration of how science and faith can be regarded as integrated, not disparate systems. For his work, Rabbi Sacks was awarded the Templeton Award in 2016.

12. Simon Jacobson, *Toward a Meaningful Life: The Wisdom of the Rebbe Menachem Mendel Schneerson* (New York: Harper-Collins, 1995).

    Menachem Mendel Schneerson, better known as the Rebbe, was the beloved leader of the Lubavitcher movement. The Rebbe's entire life was devoted to invigorating the life of the soul and being a catalyst to each and every individual's connection to God, including my own. While I was writing my endnotes, this book helped me clarify my own beliefs and cultivate the sacred in my life. I am grateful to Rabbi Jacobson for writing such an essential work.

Chapter 2: Does God Exist?

1. T. S. Eliot, "Choruses from 'The Rock,'" in *The Complete Poems and Plays* (New York: Harcourt Brace, 1971), 96.

2. Iain McGilchrist, *The Master and His Emissary: The Divided Brain and the Making of the Western World* (New Haven and London: Yale University Press, 2009). For readers interested in how our brains have shaped religious thought,

language, and belief systems over time, this book is an essential read. McGilchrist is a professor of psychiatry, and his book explores the inherent dichotomy of our brains and how each hemisphere offers its own particular construct of reality.

3. Gabriel Anton, "On Focal Diseases of the Brain Which Are Not Perceived by the Patient," lecture, Society of Physicians of Styria, Austria, December 20, 1897.

4. Andrew Newberg, "Cerebral Blood Flow During Meditative Prayer: Preliminary Findings and Methodological Issues," *Perceptual and Motor Skills* 97, no. 2 (October 2003): 625–30.

5. Paul Ekman and Richard J. Davidson, eds., *Nature of Emotion: Fundamental Questions,* Series in Affective Science (New York: Oxford University Press, 1994).

6. McGilchrist, *The Master and His Emissary.*

7. Brandon Carter, "Large Number Coincidences and the Anthropic Principle in Cosmology," in *Confrontation of Cosmological Theories with Observational Data,* ed. M. S. Longair Proceedings of the Symposium, Krakow Poland, September 10–12, 1973 (Dordrecht: D. Reidel Publishing, 1974), 291–98; later published in *General Relativity and Gravitation* 43, no. 11 (November 2011): 3225–33.

8. Tim Folger, "Does the Universe Exist If We're Not Looking?," *Discover,* June 2002.

9. Ronald H. Nash, *The Light of the Mind: St. Augustine's Theory of Knowledge* (Lexington, KY: University Press of Kentucky, 1969).

10. Michael Mayne, *The Sunrise of Wonder: Letters for the Journey* (London: Fount Paperbacks / HarperCollins, 1995), 112.

11. Aldous Huxley, *The Perennial Philosophy* (New York: Harper Perennial, 2004).

12. Giacomo Rizzolatti, Leonardo Fogassi, and Vittorio Gallese, "Neurophysiological Mechanisms Underlying the Understanding and Imitation of Action," *Nature Reviews Neuroscience* 2, no. 9 (September 2001): 661–70.

Dr. Rizzolatti is a neurophysiologist and professor at the

University of Parma in Italy. He discovered mirror neurons while doing research on the neural representation of motor movements. Mirror neurons may explain why we are able to "read" other people's minds and experience empathy.

13. Simon Baron-Cohen, "Precursors to a Theory of Mind: Understanding Attention in Others," in *Natural Theories of Mind: Evolution, Development, and Simulation of Everyday Mindreading,* ed. Andrew Whiten (Oxford: Basil Blackwell, 1991), 233–51.

14. David Constantine, "Science Illustrated: They Look Alike, but There's a Little Matter of Size," *New York Times,* August 15, 2006.

Chapter 3: The Neuroscience of the Soul

1. Isaac Newton, *Mathematical Principles of Natural Philosophy,* trans. Andrew Motte, revised by Florian Cajore, Great Books of the Western World 34 (Chicago: Encyclopedia Britannica and William Benton, 1952).

   Isaac Newton was one of the greatest scientists of all time and also a man of deep faith. While he is widely known for his discoveries related to gravity and motion, Newton was steeped in religious mysticism, including Kabbalistic notions regarding the nature of the universe.

2. Daniel Dennet, *The Atheism Tapes,* part 6, BBC TV documentary presented by Jonathan Miller, produced by Richard Denton, recorded 2003, broadcast 2004.

3. Hermann Helmholtz, "Concerning the Perceptions in General," in *Treatise on Physiological Optics,* ed. James P. C. Southall, 3rd ed., vol. 3 (1866; repr., New York: Dover, 1962).

   Hermann Ludwig Ferdinand von Helmholtz was a German scientist from the nineteenth century who is credited with many important breakthroughs in science, including conservation laws in physics and biology. Helmholtz sought a basis for neuronal energy based upon the principles of energy conservation, and his insights are the foundation for Sigmund Freud's ideas about the function of the mind, and more recently those of Karl Friston. Helmholtz also developed the ophthalmoscope.

4. Karl J. Friston and Klaas E. Stephan, "Free-Energy and the Brain," *Synthese* 159, no. 3 (2007): 417–58.

Karl Friston is the scientific director of the Wellcome Trust Centre for Neuroimaging. He is one of the foremost neuroscientists in the world and is best known for his work on developing international standards for analyzing brain imaging data. Friston has also written extensively on the neuroscientific basis of Freud's original work on the biology of emotions. The brain is a self-organized system constantly working to maintain physiological homeostasis or balance. Our thoughts, beliefs, and actions are designed to reduce uncertainty and the unpredictable.

Karl J. Friston, "The Free Energy Principle: A Unified Brain Theory?," *Nature Reviews Neuroscience* 11, no. 2 (2010): 127–38.

5. K. O. Lim and J. A. Helpern, "Neuropsychiatric Applications of DTI—A Review," *NMR in Biomedicine* 15, nos. 7–8 (2002): 587–93.

DTI is an emerging brain imaging technology that assesses the movement of water in the brain. It is being increasingly utilized for traumatic brain injuries and other disorders in which traditional brain imaging studies may not reveal the extent of damage in the brain.

6. Wilder Penfield, *The Mystery of the Mind: A Critical Study of Consciousness and the Human Brain*, Princeton Legacy Library (Princeton, NJ: Princeton University Press, 1978), 80–81.

7. Ibid.

8. David Bohm, *Wholeness and the Implicate Order* (London: Routledge and Kegan Paul, 1980).

9. G. W. F. Hegel, *Phenomenology of Spirit*, trans. A. V. Miller (Oxford: Oxford University Press, 1977), 420.

10. "God's light is the soul of man," Proverbs 20:27.

11. Adin Steinsaltz, *The Thirteen Petalled Rose: A Discourse on the Essence of Jewish Existence and Belief* (New York: Basic Books, 1980).

Chapter 4: The Evolution of Faith and Reason

1. Ludwig Wittgenstein, *Tractatus Logico-Philosophicus* (New York: Harcourt Brace and Company, 1922).

2. *The Martin Buber Reader: Essential Writings*, ed. Asher Biemann (New York: Palgrave Macmillan, 2002).

3. Adam Cohen, "Can Animal Rights Go Too Far?" *Time*, July 14, 2010, http://content.time.com/time/nation/article/0,8599,2003682,00.html.

4. Rachel Hartigan Shea, "Q&A: Pets Are Becoming People, Legally Speaking," *National Geographic*, April 7, 2014, http://news.nationalgeographic.com/news/2014/04/140406-pets-cats-dogs-animal-rights-citizen-canine/.

5. Charles Darwin, *The Descent of Man* (London: John Murray, 1871), 1:168.

6. Tamara B. Franklin et al., "Epigenetic Transmission of the Impact of Early Stress Across Generations," *Biological Psychiatry* 68, no. 5 (2010): 408–15.

7. McGilchrist, *The Master and His Emissary.*

8. *The Selected Poetry of Rainer Maria Rilke*, trans. Stephen Mitchell (New York: Vintage, 1982).

9. Stuart Kauffman, *Reinventing the Sacred: A New View of Science, Reason, and Religion* (New York: Basic Books, 2008), 129–30.

10. Sacks, *The Great Partnership.*

11. The expression "barbaric yawp" comes from Walt Whitman's work "Song of Myself," in *Leaves of Grass* (1855; repr. Philadelphia: Sherman & Co., 1990).

12. Exodus 4:10.

13. John Milton, *Paradise Lost*, bk. 7, lines 176–79.

14. Ibid., bk. 1, lines 254–55.

15. Martin Buber, "The Faith of Judaism," in *The Martin Buber Reader: Essential Writings*, ed. Asher D. Biemann (New York: Palgrave Macmillan, 2002), 99.

16. Ibid.

Chapter 5: What's the Meaning of Life?

1. Sacks, *The Great Partnership*, 37.

2. Viktor Frankl, *Man's Search for Meaning*. Originally published 1946.

3. Paul Bignell, "42: The Answer to Life, the Universe and Everything," *Independent*, February 5, 2011, http://www.independent.co.uk/life-style/history/42-the-answer-to-life-the-universe-and-everything-2205734.html.

4. Erwin Schrodinger, *What Is Life? With Mind and Matter* (Cambridge, UK: Cambridge University Press, 1944).

5. Gerald Edelman, *Neural Darwinism: The Theory of Neuronal Group Selection* (New York: Basic Books, 1987).

6. Martin H. Teicher et al., "Childhood Neglect Is Associated with Reduced Corpus Callosum Area," *Biological Psychiatry* 56, no. 2 (2004): 80–85.

7. Sigmund Freud, *Beyond the Pleasure Principle*, in *The Essentials of Psycho-Analysis: The Definitive Collection of Sigmund Freud's Writing*, trans. James Strachey (London: Penguin Books, 1991).

    Despite the common misconception that Freud mired himself completely in resurrected Greek myths, the fact is that he constructed a model of the human mind in terms of biological forces that has been validated by neuroscience. These include his notions of the "death wish," which describes the force of self-annihilation within the brain and which has been actively demonstrated as a process now referred to as apoptosis.

8. The Nobel Prize in Physiology or Medicine was awarded jointly to Sydney Brenner, H. Robert Horvitz, and John E. Sulston in 2002 "for their discoveries concerning genetic regulation of organ development and programmed cell death,'" Nobelprize.org.

9. Andrew G. Renehan, Catherine Booth, and Christopher S. Potten, "What Is Apoptosis, and Why Is It Important?," *British Medical Journal* 322, no. 7301 (2001): 1536–38, http://www.ncbi.nlm.nih.gov/pmc/articles/PMC1120576.

10. J. J. Miguel-Hidalgo et al., "Apoptosis-Related Proteins and Proliferation Markers in the Orbitofrontal Cortex in

Major Depressive Disorder," *Journal of Affective Disorders* 158 (2014): 62–70.

11. Attributed to Mahatma Gandhi, quoted in Taro Gold, *Open Your Mind, Open Your Life: A Little Book of Eastern Wisdom* (Kansas City, MO: Andrews McMeel Publishing, 2001).

12. "Heart Disease: It's Partly in Your Head," *Harvard Heart Letter*, March 2014, Harvard Health Publications.

13. Salim S. Virani et al., "Takotsubo Cardiomyopathy, or Broken-Heart Syndrome," *Texas Heart Institute Journal* 34, no. 1 (2007): 76–79.

14. Blaise Pascal, *Pensées,* trans. A. J. Krailsheimer (London: Penguin Books, 1966), 423.

15. Maimonides, Code of Law, Laws of Repentance 3.4.

Chapter 6: Are We Free?

1. Arthur Schopenhauer, *The World as Will and Representation*.

2. Yosef Y. Jacobson, "On the Essence of Freedom," Weekly Torah, Kabbalah Online, http://www.chabad .org/kabbalah/article_cdo/aid/3182099/jewish/On-the -Essence-of-Freedom.htm.

3. Hippocrates, *The Genuine Works of Hippocrates*, trans. Francis Adams (London: Sydenham Society, 1849), 2:344–45.

4. Jean-Paul Sartre, *Existentialism Is a Humanism*, trans. Carol Macomber (New Haven, CT: Yale University Press, 2007).

   This important book, first published in 1946, is about the perspective of existentialism and our inherent ability to self-create.

5. Walter Isaacson, *Einstein: His Life and Universe* (New York: Simon and Schuster, 2007).

6. Rabbi Daniel Lapin, *Thought Tools*, no. 16, April 17, 2008, http://www.rabbidaniellapin.com/thoughttools/ TT1641708.pdf.

7. Benjamin Libet, Curtis A. Gleason, Elwood W. Wright, and Dennis K. Pearl, "Time of Conscious Intention to Act in Relation to Onset of Cerebral Activity (Readiness Potential): The Unconscious Initiation of a Freely Voluntary Act," *Brain* 106, pt. 3 (1983): 623–42.

In the early 1980s, Libet found that a readiness potential (RP) over central scalp locations begins, on average,
several hundred milliseconds before the reported time of
awareness of willing to move (W). Patrick Haggard and
Martin Eimer later found no correlation between the timing of the RP and W, suggesting that the RP does not reflect processes causal of W. This conclusion has significant
implications for arguments about free will and the causal
role of consciousness. (Patrick Haggard and Martin Eimer,
"On the Relation Between Brain Potentials and the Awareness of Voluntary Movements," *Experimental Brain Research*
126, no. 1 [1999]: 128–33.)

8. William James, "The Dilemma of Determinism," in *Philosophers of Process*, eds. Douglas Browning and William T.
Myers (New York: Fordham University Press, 1998).

9. Malcom Gladwell, *Blink: The Power of Thinking Without
Thinking* (New York: Little, Brown, 2007), 58.

10. S. Pockett, "On Subjective Back-Referral and How Long It
Takes to Become Conscious of a Stimulus: A Reinterpreta-
tion of Libet's Data," *Consciousness and Cognition* 11, no. 2
(2002): 144–61.

The original data reported by Benjamin Libet and colleagues are reinterpreted, taking into account the facili-
tation that is experimentally demonstrated in the first of
their series of articles. It is shown that the original data
equally well or better support a quite different set of conclusions from those drawn by Libet. The new conclusions
are that it takes only 80 ms (rather than 500 ms) for stimuli to come to consciousness, and that "subjective back-
referral of sensations in time" to the time of the stimulus
does not occur (contrary to Libet's original interpretation
of his results).

11. Angus Menuge, "Does Neuroscience Undermine Retribu-
tive Justice?," in *Free Will in Criminal Law and Procedure,* ed.
Friedrich Toepel, Proceedings of the 23rd and 24th IVR
World Congress, Kraków 2007 and Beijing 2009 (Stutt-
gart: Franz Steiner Verlag, 2010), 73–94.

12. Massimo Pigliucci, "Is Science All You Need?," *Philosophers'*

Magazine 57 (2nd quarter 2012): 111–12, http://philpapers
.org/archive/PIGISA.

13. Itzhak Fried, "Internally Generated Preactivation of Single Neurons in Human Medial Frontal Cortex Predicts Volition," *Neuron* 69, no. 3 (2011): 548–62.

14. Richard Dawkins, *The God Delusion* (New York: Houghton Mifflin, 2006), 213.

   Dawkins discusses evolution and rejects the notion that anything exists that is "irreducibly complex." An avowed atheist, Dawkins believes that our existence is an infinite regress of materialism: "It's turtles all the way down." But what Dawkins is unable to sweep under the rug is human consciousness, where the rules of reductionism do not apply. The biological and physical mechanisms of consciousness have eluded a naturalistic explanation.

15. John C. Eccles, *Evolution of the Brain: Creation of the Self* (Oxford: Taylor & Francis, 1991).

16. Martha Stout, *The Sociopath Next Door* (New York: Broadway Books, 2005).

17. Simon Baron-Cohen, "Precursors to a Theory of Mind: Understanding Attention in Others," in *Natural Theories of Mind: Evolution, Development and Simulation of Everyday Mindreading,* ed. Andrew Whiten (Oxford: Basil Blackwell, 1991), 233–51; Simon Baron-Cohen, "Autism: A Specific Cognitive Disorder of 'Mind-Blindness,'" *International Review of Psychiatry* 2, no. 1 (1990): 81–90.

   Simon Baron-Cohen is a professor of developmental psychopathology at the University of Cambridge and the director of the Autism Research Centre. Baron-Cohen theorized that autism is a theory-of-mind impairment. We all possess a "theory of mind" because the minds of other people are irreducible, and not directly observable, and because we have no direct way of verifying that other people even have minds at all; we can only ever implicitly infer the existence of minds in other people based upon our own minds and experience.

18. Naoki Higashida, *The Reason I Jump: The Inner Voice of a*

Thirteen-Year-Old Boy with Autism, trans. K. A. Yoshida and David Mitchell (New York: Random House, 2013).

19. Emiliano Ricciardi et al., "How the Brain Heals Emotional Wounds: The Functional Neuroanatomy of Forgiveness," *Frontiers in Human Neuroscience* 7 (2013): 839.

20. Hannah Arendt, *The Human Condition* (Chicago: University Chicago Press, 1958), 241.

Arendt is best known for her expression "the banality of evil." I included her thoughts because if she were alive today, she would agree with Naoki Higashida's explanation of the world and human behavior.

Chapter 7: Do Good and Evil Actually Exist?

1. Adin Steinsaltz, *Thirteen Petalled Rose,* 96.

Many of my ideas about the nature of the soul were inspired by this book. Steinsaltz is a foremost scholar and has authored numerous books on the subject of Jewish thought.

2. J. Krishnamurti, *Freedom from the Known* (New York: Harper and Row, 1969), 13.

3. American Psychiatric Association, *Diagnostic and Statistical Manual of Mental Disorders,* 5th ed. (Washington, DC: American Psychiatric Association, 2013).

4. Daniel Goleman, "Probing the Enigma of Multiple Personality," *New York Times,* June 28, 1988, http://www.nytimes.com/1988/06/28/science/probing-the-enigma-of-multiple-personality.html.

5. Scott D. Miller, "Optical Differences in Cases of Multiple Personality Disorder," *Journal of Nervous and Mental Disease* 177, no. 8 (1989): 480–86.

6. Robert Louis Stevenson, *Dr. Jekyll and Mr. Hyde* (1886; repr. London: Bibliolis Books, 2010), 95.

7. Ibid.

8. R. C. Byrd, "Positive Therapeutic Effects of Intercessory Prayer in a Coronary Care Unit Population," *Southern Medical Journal* 81, no. 7 (1988): 826–29.

9. Giovanni Pico della Mirandola, *Heptaplus,* quoted in C. G.

Jung and W. E. Pauli, *The Interpretation of Nature and the Psyche* (New Jersey: Princeton University Press, 1975).

10. Sharon Dekel, Zahava Solomon, and Eyal Rozenstreich, "Secondary Salutogenic Effects in Veterans Whose Parents Were Holocaust Survivors?," *Journal of Psychiatric Research* 47, no. 2 (2013): 266–71.

11. Henry Melvill, "Partaking in Other Men's Sins," address to St. Margaret's Church, Lothbury, England, June 12, 1855, printed in *The Golden Lectures* (London: James Paul, 1855).

12. Masahiro Kawasaki et al., "Inter-Brain Synchronization During Coordination of Speech Rhythm in Human-to-Human Social Interaction," *Scientific Reports* 3, no. 1692 (2013); Uri Hasson et al., "Brain-to-Brain Coupling: A Mechanism for Creating and Sharing a Social World," *Trends in Cognitive Sciences* 16, no. 2 (2012): 114–21.

13. Allan N. Schore, *Affect Regulation and the Repair of the Self*, Norton Series on Interpersonal Neurobiology (New York: W. W. Norton, 2003).

14. Sara Laskow, "The Role of the Supernatural in the Discovery of EEGs," *The Atlantic*, November 23, 2014.

    This is a good story about Hans Berger, who developed the first device to measure brain waves. Is it no coincidence that Berger also believed in telepathy?

15. Ibid.

16. Donald Capps, *The Religious Life: The Insights of William James* (Eugene, OR: Cascade Books, 2015).

17. Philologus, "Roots of 'Religion,'" *The Forward*, May 25, 2007.

18. Exodus 3:14.

19. Avivah Gottlieb Zornberg, *The Particulars of Rapture: Reflections on Exodus* (New York: Doubleday, 2001), 75.

    Avivah Zornberg offers a beautiful and elegant interpretation of God's mysterious reply to Moses's inquiry about God's name, based upon the original writings and teachings of the Maharal of Prague.

Chapter 8: Immortality: The Remembrance of What Is

1. T. S. Eliot, "Little Gidding," *Four Quartets,* in *The Complete Poems and Plays, 1909–1950* (New York: Harcourt Brace, 1971), 145.

2. Milan Kundera, *The Unbearable Lightness of Being* (1984; repr. New York: Perennial Classics, 1999).

3. Ewald Hering, *On Memory and the Specific Energies of the Nervous System* (Chicago: Open Court Publishing Company, 1895).

4. William Butler Yeats, *The Second Coming* (Ireland: The Cuala Press, 1921).

5. Karl Pribram, *Rethinking Neural Networks: Quantum Fields and Biological Data* (New York: Lawrence Erlbaum Associates, 1993).

6. Julian B. Barbour, *The End of Time: The Next Revolution in Physics* (London: Weidenfeld & Nicolson, 1999).

7. Talmud Tractate Berachot 57b.

8. Freud, *Dream Psychology*, 172.

9. William James, *The Varieties of Religious Experience: A Study in Human Nature,* The Gifford Lectures on Natural Religion Delivered at Edinburgh 1901–1902 (New York: Modern Library / Random House, 1902), 490.

10. Psalms 126:1.

11. Max Scheler, *On the Eternal in Man* (1960; repr. New Brunswick, NJ: Transaction Publishers, 2009).

    Max Scheler was an important scholar on the philosophy of religion, and Pope John Paul II was greatly influenced by his works.

12. Jean-Paul Sartre, *The Emotions Outline of a Theory,* trans. Bernard Frechtman (New York: Philosophical Library, 1948).

13. William Wordsworth, "Lines Written a Few Miles Above Tintern Abbey," *Lyrical Ballads* (1798; repr. London: Penguin Books, 2006), 112.

14. Sigmund Freud, "The Project for a Scientific Psychology" in *The Standard Edition of the Complete Psychological Works of Sigmund Freud,* vol. 16, ed. James Strachey (London, Hogarth Press, 1963).

15. Thubten Chodron, *Buddhism for Beginners* (Boston: Snow Lion, 2001).

16. "Pope Francis to Religious: Do Not Fall into Spiritual Alzheimer's," Rome Reports news agency, July 9, 2015.

17. Don DeLillo, *Cosmopolis: A Novel* (New York: Scribner, 2003), 155.

18. Pascal, *Pensées*.

19. Rabbi Isaac Luria, 1534–1572, the father of modern Kabbalah.

20. Erwin Schrödinger, *What Is Life?* (Cambridge: Cambridge University Press, 1967), 21.

21. The saying "Remembrance leads to redemption" is attributed to the Baal Shem Tov and inscribed on the gates of the Yad Vashem Holocaust History Museum.

Acknowledgments

The path to this book's creation began with several passionate discussions I had with Celeste Fine, my literary agent at Sterling Lord Literistic (SLL). It was through Celeste's incredible inquisitiveness and openness to new ideas that this book was born. I remember receiving her best compliment—that the original draft of this manuscript reminded her of her grandfather's attempt to put his words about faith on hundreds of separate notes. I hope I would not have disappointed him in my attempt to do the same. Thanks also go to agent John Maas at SLL for his championing of this project and his continued nurturance of it.

I first met Gary Jansen, my editor at Crown, an imprint of Penguin Random House, over lunch at a Cuban restaurant in Midtown Manhattan on a cold winter Monday in 2013. We spoke for several hours about our mutual interest in science and religion, our deep appreciation of William James, and in discovering over and over that we shared an excitement for these two basic frameworks of all human understanding (science and religion), we both sought in our lives to aim at reconciling these seemingly disparate worldviews for the betterment of all. At every step in this process, Gary has challenged me to draw from a deeper well of insight, while respecting the core message that has always been its original intent. His tendering of this work has enriched it greatly.

Editors Ian Blake Newhem and Marcus Brotherton helped

shape the manuscript along the way and added clarity to the discussion and insight into how to broaden this message to its fullest. Both are extraordinary individuals and great writers, and without their help this book would never have materialized. Ian is like a poet who has helped me see the elegance of language in nature and nature in language. A special call-out goes to Marcus, who has helped me grow so deeply in the process of writing this book, leading me to discover deeper aspects of faith and meaning within myself.

I dare say that *miraculously* Pope Francis has proved an unexpected inspiration during the year of writing this book. He's challenging all of us, regardless of apparent differences in faith—or the absence of articulated faith—to be better humans, whether through the pursuit of religion, science, or philosophy. His masterpiece, *The Name of God Is Mercy*, is the most important message of our lifetime, regardless of our religious persuasion.

I'm honored to know so many people in my life who have been instrumental to my learning, both before and during the unfolding of this book. First, my family, and especially my wife, Rita, who lives every day with a level of kindness and love that I am blessed to have in my life, and my daughters, Julia and Sofia, are both relentless in their questions about the meaning of science, religion, and life. It's through their existentially wealthy questions, and my best attempts to answer them, that most of my ideas about *relationships* and *our connection to God* were born. And it's through their love that I have come to learn of the reality of God's existence. Both my parents are now in the "World of Understanding" but always taught by way of example the importance of love and family. The closeness of my family, and especially to my sister Joanne, brother Robert, Scott Ross, cousins Ken, Marion, Warren, and Steve, my nephews, nieces, and grandnieces Blanca, Raul, Cathy, Lou, and Bea, are my personal examples of how faith is materialized.

From a religious and philosophical perspective, I am deeply indebted to the teachings of the *Lubavitcher Rebbe of Blessed Memory* as they were taught to me by my most special and dearest friend, Rabbi Noson Gurary. I love Rabbi Gurary, a truly righteous person I am deeply blessed to have in my life. His faith is an inspiration to all who know him.

I am also blessed to have so many dear friends, as true and loyal as one could ever have, including my most enduring friendships with Sammy Fox, Dr. Bert Pepper—a sage in the true meaning of the word and who has shown me such kindness and generosity of spirit in a difficult time in my life—Tim Flaherety, Dan Kolak, Perry Bard, Larry Zaret, Michael Shulman, Madeline Sorrentino, Steve Kraskow, Mitch Golden, Thom Gencarelli, Kathleen O' Connell, Jay and Brenda Lender, Robin Matza, Peter Kash, Linda Friedland, Peter Madill, Carl Germano, Kamran Fallapour, James Gordon, Aron Weber, David Ober, Michael Kelly, Dr. Chris Renna, and Harvey Weinstein. I am so blessed to have these individuals in my life and am grateful for their love and kindness.

I am also deeply indebted and grateful to have work-related friendships that go beyond work because of our shared vision: Dr. Ron Dozoretz, Robert Gibbs, Nancy Grden, Laura Miles, Dr. Rudy Tanzi, Patrick Kennedy, Dr. Lloyd Sederer, Darnley Stewart, Dr. Chris Renna, Dr. Jonathan Mann, and Dr. Duane Mitchell, and Dr. Dan Lucksaker.

Maimonides, Spinoza, Martin Buber, William James, Sigmund Freud, Hanna Arendt, Iain McGilchrist, Rabbi Jonathan Sacks, Avivah Zornberg, Adin Steinsaltz, Jean-Paul Sartre, and Dr. Dan Kolak are some of the intellectual giants and wellsprings of wisdom from which I have often drunk.

There were many summer days when I sat at my computer overlooking majestic Lake George, and I want to thank the owners and staff of Canoe Island Lodge for being such wonderful soul attendants during the writing of sections of this book.

Finally, I am grateful to those of you who've embarked on reading this work, which has been not only a labor of love, but the culmination of a life's interrogations. It's difficult in that milieu for us to know what we should pay attention to, and it's my loftiest hope that you find the message of this book personally relevant and meaningful.

With deepest gratitude,

Dr. Jay Lombard
New York City, July 2016

Dr. Jay Lombard is a neurologist and cofounder of Genomind, a precision medicine company for patients with psychiatric and neurological conditions.

Prior to cofounding Genomind, Dr. Lombard served as chief of neurology at Bronx Lebanon Hospital between 2007 and 2011. He has held academic medical appointments at New York Presbyterian Hospital and Albert Einstein College of Medicine.

Dr. Lombard served as chief of neurology at Westchester Square Medical Center from 1995 to 2000, and at the Brain Behavior Center in Rockland County, New York, from 2000 to 2007, where he specialized in patients with intractable neurological and psychiatric disorders, including Alzheimer's disease, ALS, and autism. He was honored as one of the top neurologists in New York.

Dr. Lombard has written extensively on topics in psychiatry and neurology, including several peer-reviewed papers and medical textbooks, among them the *New England Journal of Medicine, Medical Hypotheses, Clinics of North America, American Journal of Managed Care,* and *Expert Opinion.* His research has included an original hypothesis linking autism to mitochondrial dysfunction, which has subsequently been validated by recent genetic findings. He has also published articles in peer-reviewed medical journals covering diverse topics from Alzheimer's disease, pediatric neuropsychiatric disorders, and pharmacogenomics in psychiatry.

Dr. Lombard is also a widely acclaimed author of popular nonfiction works related to the effects of nutrition and the brain, including the *Brain Wellness Plan*, which he coauthored with Carl Germano (Kensington 1998); *Balance Your Brain, Balance Your Life*, coauthored with Dr. Chris Renna (Wiley 2004); and *Freedom from Disease*, coauthored with Peter Kash (St. Martin's 2009).

He lectures on these topics frequently, both nationally and internationally.

Dr. Lombard has also been an adviser to the film industry, where he served as a medical consultant for Academy Award–winning directors Jonathan Demme and Martin Scorsese on Hollywood feature films. He has also appeared as a guest on numerous TV and radio programs, including *Larry King Live*, CBS News, the Food Network, and with Dr. Mehmet Oz, to discuss the challenges and advances in treating intractable neurological conditions.

In April 2012, Dr. Lombard appeared as a featured speaker at *TedMed*, where he lectured on the challenges of understanding complex disorders of the brain. Jay Walker, founder of *TedMed*, described Dr. Lombard as "part Freud, part Sherlock Holmes."

Dr. Lombard completed his postdoctoral neurology training at Long Island Jewish Medical Center in 1994.

He maintains a consultative practice in New York City, specializing in neurobehavioral conditions.

Printed in the United States
by Baker & Taylor Publisher Services